Cambridge Tracts in Mathematics and Mathematical Physics

No. 16

LINEAR ALGEBRAS

LINEAR ALGEBRAS

by

L. E. DICKSON, Ph.D.
Professor of Mathematics in the University of Chicago

Cambridge:
at the University Press
1914

CAMBRIDGE
UNIVERSITY PRESS

University Printing House, Cambridge CB2 8BS, United Kingdom

Cambridge University Press is part of the University of Cambridge.

It furthers the University's mission by disseminating knowledge in the pursuit of education, learning and research at the highest international levels of excellence.

www.cambridge.org
Information on this title: www.cambridge.org/9781107493940

First published 1914
Re-issued 2015

A catalogue record for this publication is available from the British Library

ISBN 978-1-107-49394-0 Paperback

PREFACE

THE theory of linear associative algebras (or closed systems of hypercomplex numbers) is essentially the theory of pairs of reciprocal linear groups (§ 52) or the theory of certain sets of matrices or bilinear forms (§ 53). Beginning with Hamilton's discovery of quaternions seventy years ago, there has been a rapidly increasing number of papers on these various theories. The French Encyclopedia of Mathematics devotes more than a hundred pages to references and statements of results on this subject (with an additional part on ordinary complex numbers). However, the subject is rich not merely in extent, but also in depth, reaching to the very heart of modern algebra.

The purpose of this tract is to afford an elementary introduction to the general theory of linear algebras, including also non-associative algebras. It retains the character of a set of lectures delivered at the University of Chicago in the Spring Quarter of 1913. The subject is presented from the standpoint of linear algebras and makes no use either of the terminology or of theorems peculiar to the theory of bilinear forms, matrices or groups (aside of course from §§ 52–54, which treat in ample detail of the relations of linear algebras to those topics).

Part I relates to definitions, concrete illustrations, and important theorems capable of brief and elementary proof. A very elementary proof is given of Frobenius's theorem which shows the unique place of quaternions among algebras. The remarkable properties of Cayley's algebra of eight units are here obtained for the first time in a simple manner, without computations. Other new results and new points of view will be found in this introductory part.

In presenting in Parts II and IV the main theorems of the general theory, it was necessary to choose between the expositions by Molien, Cartan and Wedderburn (that by Frobenius being based upon bilinear forms and hence outside our plan of treatment). We have not presented the theory of Molien partly because his later proofs depend upon the theory of groups and partly because certain of his earlier proofs have not yet been made correctly by his methods. The more general paper by Wedderburn is based upon a rather abstract calculus of complexes, comparable with the theory of abstract groups (§ 61). In compensation, he obtains in relatively brief space the main theorems not only for the usual cases of complex and real algebras, but also for algebras the coordinates of whose numbers range over any field (stated in § 56).

In order that our treatment of the general theory shall be elementary and concrete and shall use but a very few concepts easily kept in mind, we have confined our exposition (in Part II) to the classical case of algebras whose numbers have ordinary complex coordinates and given a careful revision of the theory as presented in Cartan's fundamental paper. Running parallel with the general theory is an illustrative example treated independently but in the spirit of the theory. While we thereby lose the generalization to a general field, we gain the important normalized sets of units, first given by Scheffers under certain restrictions, and so obtain the analogues of important theorems on the canonical forms of groups of linear transformations or of sets of matrices or bilinear forms.

I am much indebted to Professor Wedderburn of Princeton and Miss Hazlett of Chicago for suggestions made after careful readings of the proofs. My thanks are due to the editors for the opportunity to participate in this useful series of tracts. Finally, I am under obligations to the officials of the University Press for complying with all of my requests as regards the form of this tract, and for expeditious publication in spite of the distance travelled by the proofs. The quality of the printing speaks for itself.

L. E. D.

CHICAGO,
May, 1914.

CONTENTS

Part III. Relations of Linear Algebras to other Subjects

Part IV. Linear Algebras over a Field F

PART I

DEFINITIONS, ILLUSTRATIONS AND ELEMENTARY THEOREMS

1. Arithmetical definition of ordinary complex numbers. The following purely arithmetical theory of couples (a, b) of real numbers differs only in unessential points from the initial theory of W. R. Hamilton*. Two couples (a, b) and (c, d) are called equal if and only if $a = c$, $b = d$. Addition, subtraction and multiplication of two couples are defined by the formulas†

$$\left.\begin{aligned}(a, b) + (c, d) &= (a + c,\ b + d)\\ (a, b) - (c, d) &= (a - c,\ b - d)\\ (a, b)\,(c, d) &= (ac - bd,\ ad + bc)\end{aligned}\right\} \qquad (1).$$

Addition is seen to be commutative and associative:

$$x + x' = x' + x, \quad (x + x') + x'' = x + (x' + x'') \qquad (2),$$

where x, x', x'' are any couples. Multiplication is commutative, associative and distributive:

$$xx' = x'x, \quad (xx')\,x'' = x\,(x'x'') \qquad (3),$$

$$x\,(x' + x'') = xx' + xx'', \quad (x' + x'')\,x = x'x + x''x \qquad (4).$$

* *Trans. Irish Acad.*, vol. 17 (1837), p. 293; *Lectures on Quaternions*, 1853, Preface.

† Each couple (a, b) uniquely determines a vector from the origin O to the point A with the rectangular coordinates a, b. The sum of two vectors from O to A and the point $C = (c, d)$ is defined to be the vector from O to the fourth vertex S of the parallelogram having the lines OA and OC as two sides. The coordinates of S are $a + c$, $b + d$. Subtraction of vectors is the operation inverse to addition; thus $OS - OA = OC$. To define the product of the vectors from O to A and C, we employ initially the polar coordinates r, θ and r', θ' of A and C. Then $OA \,.\, OC$ is defined to be the vector from O to the point P with the polar coordinates rr', $\theta + \theta'$. Since A has the rectangular coordinates $a = r\cos\theta$, $b = r\sin\theta$, and similarly for C and P, the expansions of $\cos(\theta + \theta')$ and $\sin(\theta + \theta')$ lead to the third relation (1) between the rectangular coordinates of A, C, P.

Division is defined as the operation inverse to multiplication. Division except by (0, 0) is possible and unique:

$$\frac{(c, d)}{(a, b)} = \left(\frac{ac+bd}{a^2+b^2}, \frac{ad-bc}{a^2+b^2}\right) \qquad (5).$$

In particular, we have

$$(a, 0) \pm (c, 0) = (a \pm c, 0), \quad (a, 0)(c, 0) = (ac, 0),$$

$$\frac{(c, 0)}{(a, 0)} = \left(\frac{c}{a}, 0\right) \quad \text{if } a \neq 0.$$

Hence the couples $(a, 0)$ combine under the above defined addition, multiplication, etc., exactly as the real numbers a combine under ordinary addition, multiplication, etc. Without introducing any contradiction, we may and shall impose upon our system of couples (a, b), subject to the above definitions of addition, etc., the further assumption* that the couple $(a, 0)$ shall be the real number a. For brevity write i for (0, 1). Then

$$i^2 = (0, 1)(0, 1) = (-1, 0) = -1,$$

$$(a, b) = (a, 0) + (0, b) = a + (b, 0)(0, 1) = a + bi.$$

The resulting symbol $a + bi$ is called a complex number. Relations (1) and (5) now take the familiar forms

$$\left.\begin{aligned} (a+bi) \pm (c+di) &= (a \pm c) + (b \pm d)\, i \\ (a+bi)(c+di) &= (ac - bd) + (ad + bc)\, i \\ \frac{c+di}{a+bi} &= \frac{ac+bd}{a^2+b^2} + \frac{ad-bc}{a^2+b^2}\, i \end{aligned}\right\} \qquad (6),$$

where, for the last, $a + bi \neq 0$, i.e. a and b are not both zero.

2. Number fields. A set of complex numbers is called a number field (domain of rationality or Körper) if the rational operations can always be performed unambiguously within the set. In other words, the sum, difference, product and quotient (the divisor not being zero) of any two equal or distinct numbers of the set must be numbers belonging to the set.

In view of (6), all complex numbers $a + bi$ form a field. Again, all real numbers form a field. The set of all rational numbers is a field, but the set of all integers is not.

* Just as the natural numbers are included among the signed integers, the integers among the rational numbers, and the latter among the real numbers defined by means of them. In the same train of ideas, 1 is often used to denote the principal unit (§ 7, § 11), and the number e for the scalar matrix S_e (§ 4, second foot-note).

3. Matrices. The concept matrix* affords an excellent introduction to the subject of this tract and, moreover, is of special importance in the general theory. We shall consider only square matrices of n rows each containing n elements. For example, if $n=2$,

$$m=\begin{pmatrix} a & b \\ c & d \end{pmatrix}, \quad \mu=\begin{pmatrix} \alpha & \beta \\ \gamma & \delta \end{pmatrix} \tag{7}$$

are matrices, the elements of the first matrix m being a, b, c, d. Each element may be any number of a given number field F. We shall say that m and μ are equal if and only if their corresponding elements are equal, $a=\alpha$, etc. Addition and multiplication are defined by

$$m+\mu=\begin{pmatrix} a+\alpha & b+\beta \\ c+\gamma & d+\delta \end{pmatrix}, \quad m\mu=\begin{pmatrix} a\alpha+b\gamma & a\beta+b\delta \\ c\alpha+d\gamma & c\beta+d\delta \end{pmatrix} \tag{8}.$$

The element in the ith row and jth column of the product is the sum of the products of each element of the ith row of the first matrix by the corresponding element of the jth column of the second matrix, i.e. first by first, second by second, etc. This rule holds also for matrices of n^2 elements. Of the four possible rules for expressing the product of two determinants of order n as a determinant of order n, the above is the only rule which holds also for matrices.

With the exception of (3_1), the laws (2)—(4) for addition and multiplication hold for matrices. Since the product (8_2) is in general altered when the Roman and Greek letters are interchanged, matric multiplication is usually not commutative. Accordingly we shall see that we must distinguish between two distinct kinds of division of matrices. To this end, note that

$$\begin{pmatrix} e & 0 \\ 0 & e \end{pmatrix} m = m \begin{pmatrix} e & 0 \\ 0 & e \end{pmatrix} = \begin{pmatrix} ea & eb \\ ec & ed \end{pmatrix} \tag{9}.$$

In particular, the *unit* matrix

$$I=\begin{pmatrix} 1 & 0 \\ 0 & 1 \end{pmatrix} \tag{10}$$

has the property that $Im=mI=m$, for every matrix m. By the *inverse* of a matrix m whose determinant $\Delta=|m|$ is not zero is meant

* Cayley, *Phil. Trans. London*, vol. 148 (1858), pp. 17—37 (=*Coll. Math. Papers*, II, pp. 475, 604).

$$m^{-1}=\begin{pmatrix}\frac{d}{\Delta} & \frac{-b}{\Delta}\\ \frac{-c}{\Delta} & \frac{a}{\Delta}\end{pmatrix} \qquad (11),$$

if $n=2$, while if $n \gtrless 2$, we employ as the element in the ith row and jth column the quotient of the co-factor of the element in the jth row and ith column of Δ by Δ. Then

$$mm^{-1}=m^{-1}m=I \qquad (12).$$

Given two matrices m and p such that $|m| \neq 0$, we can find one and only one matrix $\mu=m^{-1}p$ such that $m\mu=p$, also one and only one matrix $\nu=pm^{-1}$ such that $\nu m=p$. These respective kinds of division by p by m shall be called *right-hand and left-hand division.*

On the contrary, if $|m|=0$, there is no matrix μ for which $m\mu=I$, since this would imply $0\,|\mu|=|I|=1$. Likewise, there is no matrix ν for which $\nu m=I$.

Thus right- and left-hand division by m are each always possible and unique if and only if the determinant of m is not zero.

Addition, subtraction, multiplication and division of matrices with elements in a field F lead to matrices with elements in F. Accordingly we shall speak of the matric algebra over the field F. When F is the field of all complex numbers, the field of all real numbers, or that of all rational numbers, we have the complex, real or rational matric algebra of square matrices of n^2 elements.

4. A matric algebra viewed as a linear algebra*. Taking $n=2$, we shall make use of the particular matrices

$$e_{11}=\begin{pmatrix}1&0\\0&0\end{pmatrix},\quad e_{12}=\begin{pmatrix}0&1\\0&0\end{pmatrix},\quad e_{21}=\begin{pmatrix}0&0\\1&0\end{pmatrix},\quad e_{22}=\begin{pmatrix}0&0\\0&1\end{pmatrix} \qquad (13).$$

Their sixteen products by twos are

$$e_{ij}e_{jk}=e_{ik},\quad e_{ij}e_{tk}=0 \quad (t \neq j) \qquad (14).$$

If m is a matrix and e is a number, we shall define the product† em

* For references, see § 13.

† In the product (9) we may therefore replace the "scalar matrix" $\begin{pmatrix}e&0\\0&e\end{pmatrix}=S_e$ by the number e. This becomes intuitive if we note that $S_e=eI$. Since $S_e+S_f=S_{e+f}$, $S_eS_f=S_{ef}$, etc., the algebra of all scalar matrices over a field F is abstractly identical with F. This replacement of S_e by e is similar to that of $(a, 0)$ by a in § 1.

or me to be the matrix each of whose elements is the product of e by the corresponding element of m:

$$e\begin{pmatrix} a & b \\ c & d \end{pmatrix} = \begin{pmatrix} a & b \\ c & d \end{pmatrix} e = \begin{pmatrix} ea & eb \\ ec & ed \end{pmatrix} \qquad (15).$$

In view of (13), matrices (7) and (8) may be expressed in the form

$$\left.\begin{aligned} m &= ae_{11} + be_{12} + ce_{21} + de_{22} \\ \mu &= \alpha e_{11} + \beta e_{12} + \gamma e_{21} + \delta e_{22} \end{aligned}\right\} \qquad (16),$$

$$\left.\begin{aligned} m + \mu &= (a+\alpha)\,e_{11} + (b+\beta)\,e_{12} + (c+\gamma)\,e_{21} + (d+\delta)\,e_{22} \\ m\mu &= (a\alpha + b\gamma)\,e_{11} + (a\beta + b\delta)\,e_{12} + (c\alpha + d\gamma)\,e_{21} + (c\beta + d\delta)\,e_{22} \end{aligned}\right\} \qquad (17).$$

The last may also be found from (16) by use of relations (14).

The set of hyper-complex numbers $ae_{11} + \ldots + de_{22}$, in which $a, \ldots, d$ range independently over a field F, and for which addition and multiplication are defined by (17), is called a linear associative algebra over F with the four units $e_{11}, \ldots, e_{22}$ subject to the multiplication table (14).

For any n, let e_{ij} be the square matrix of n^2 elements all zero except that in the ith row and jth column which is unity. Then relations (14) hold. We obtain a linear associative algebra with n^2 units e_{ij}.

5. General definition of hyper-complex numbers and linear algebras*. We shall generalize the notion of couples in § 1 and, with a change of notation, the notion of quadruples (7). Consider the set of all *n-tuples* $(x_1, \ldots, x_n)$, whose *coordinates* $x_1, \ldots, x_n$ range independently over a given number field F.

Two n-tuples are called equal if and only if their corresponding coordinates are equal.

Addition and subtraction of n-tuples are defined by

$$(x_1, \ldots, x_n) \pm (x_1', \ldots, x_n') = (x_1 \pm x_1', \ldots, x_n \pm x_n') \qquad (18).$$

The product of any number ρ of the field F and any n-tuple

$$x = (x_1, \ldots, x_n)$$

is defined to be

$$\rho x = x\rho = (\rho x_1, \ldots, \rho x_n) \qquad (19).$$

* Hamilton's *Lectures on Quaternions*, 1853, Introduction. For definitions by independent postulates, see Dickson, *Trans. Amer. Math. Soc.*, vol. 4 (1903), p. 21; vol. 6 (1905), p. 344.

The n *units* are defined to be

$$e_1 = (1, 0, \ldots, 0),\ e_2 = (0, 1, 0, \ldots, 0),\ \ldots,\ e_n = (0, \ldots, 0, 1).$$

Hence any n-tuple x can be expressed in the form

$$x = x_1 e_1 + x_2 e_2 + \ldots + x_n e_n.$$

We shall call x a *hyper-complex* number, or briefly a number. In view of the definition of equality of n-tuples, x and

$$x' = x_1' e_1 + \ldots + x_n' e_n$$

are equal if and only if $x_1 = x_1', \ldots, x_n = x_n'$. In particular, $x = 0$ implies that each $x_i = 0$. Hence the units $e_1, \ldots, e_n$ are linearly independent with respect to the field F.

It is assumed that any two such numbers x and x' can be combined by an operation called multiplication subject to the distributive laws (4):

$$xx' = \sum_{i,j=1}^{n} x_i x_j' e_i e_j,$$

and such that the product xx' is a number $\Sigma z_i e_i$ with coordinates in F. Necessary conditions for the latter property are

$$e_i e_j = \sum_{k=1}^{n} \gamma_{ijk} e_k \qquad (i, j = 1, \ldots, n\,;\ \gamma\text{'s in } F) \qquad (20).$$

These are sufficient conditions, since they imply

$$xx' = y \equiv \Sigma y_k e_k, \quad y_k = \sum_{i,j=1}^{n} x_i x_j' \gamma_{ijk} \qquad (k = 1, \ldots, n) \qquad (21).$$

Properties (18) and (19) of n-tuples give

$$x \pm x' = \sum_{i=1}^{n} (x_i \pm x_i') e_i, \quad \rho x = x\rho = \sum_{i=1}^{n} (\rho x_i) e_i \qquad (22),$$

if ρ is in F. The set of all numbers $\Sigma x_i e_i$, with coordinates in F, combined under multiplication as in (21), under addition and subtraction as in (22_1), and under multiplication by a number ρ of F as in (22_2), shall be said to form a *linear algebra* (or system of hyper-complex numbers) over the field F, with the units $e_1, \ldots, e_n$ (linearly independent with respect to F) and the multiplication table (20). The n^3 numbers γ_{ijk} are called the constants of multiplication. Neither the commutative nor the associative law of multiplication is assumed.

For example, the set of all ordinary complex numbers $a+bi$ form a binary linear algebra over the field F of all real numbers, with the units 1 and i subject to the multiplication table

$$1^2=1, \quad 1.i=i.1=i, \quad i^2=-1.$$

In this algebra (§ 1), multiplication is commutative and associative. The algebra is a field $F(i)$ and may be viewed as a unary algebra over this complex field with the single unit 1.

In § 4 we considered a linear associative algebra with four units.

6. Division. Given two numbers x and y of a linear algebra, we can determine uniquely a number x' of the algebra such that $xx'=y$ provided the n linear equations at the end of (21) are solvable uniquely for $x_1', \ldots, x_n'$ in the field F. This will be the case if and only if the determinant

$$\Delta(x) = \left| \sum_{i=1}^{n} x_i \gamma_{ijk} \right| \qquad (j, k=1, \ldots, n) \qquad (23)$$

is not zero. In that case, right-hand division by x is always possible and unique.

Similarly, there is a unique solution x' in the algebra of the equation $x'x=y$ if and only if

$$\Delta'(x) = \left| \sum_{i=1}^{n} x_i \gamma_{jik} \right| \qquad (j, k=1, \ldots, n) \qquad (24)$$

is not zero. In that case, left-hand division by x is always possible and unique.

We shall call $\Delta(x)$ the *right-hand determinant* of x, and $\Delta'(x)$ the *left-hand determinant* of x.

7. Principal unit (modulus). An algebra may contain a number

$$\epsilon = \epsilon_1 e_1 + \ldots + \epsilon_n e_n,$$

called a principal unit or modulus, such that

$$x\epsilon = \epsilon x = x \qquad \text{(for every number } x \text{ of the algebra)} \qquad (25).$$

For example, $\epsilon=1$ in the binary algebra of all complex numbers $a+bi$. Again, $\epsilon = e_{11}+e_{22}$ in the matric algebra of § 4.

There cannot be a second principal unit ϵ'. More generally, there is no new number ϵ' for which $x\epsilon'=x$ for every x. For, if so, $\epsilon\epsilon'=\epsilon$, while by (25_2) for $x=\epsilon'$, $\epsilon\epsilon'=\epsilon'$, whence $\epsilon'=\epsilon$.

Conditions (25) hold if and only if

$$e_j\epsilon = \epsilon e_j = e_j \qquad (j=1, \ldots, n) \qquad (25').$$

With Kronecker, set $\delta_{jj}=1$, $\delta_{jk}=0$ if $j\neq k$. By use of (20) and the linear independence of $e_1, \ldots, e_n$, we see that (25′) are equivalent to

$$\sum_{i=1}^{n} \epsilon_i \gamma_{jik} = \delta_{jk}, \qquad \sum_{i=1}^{n} \epsilon_i \gamma_{ijk} = \delta_{jk} \qquad (j, k = 1, \ldots, n) \qquad (26).$$

Hence $$\Delta'(\epsilon) = \Delta(\epsilon) = |\delta_{jk}| = 1 \qquad (27).$$

Hence if there exists a principal unit, neither $\Delta(x)$ nor $\Delta'(x)$ is identically zero in $x_1, \ldots, x_n$.

For the case of a linear associative algebra it is easily proved that, conversely*, when neither $\Delta(x)$ nor $\Delta'(x)$ is identically zero, the algebra has a principal unit. Indeed, there is then a number u for which neither $\Delta(u)$ nor $\Delta'(u)$ is zero. By § 6, there is a unique number ϵ of the algebra such that $u\epsilon = u$, and a uniquely determined z for which $zu = x$, where x is an arbitrary number of the algebra. Then, by the associative law,

$$x\epsilon = (zu)\epsilon = z(u\epsilon) = zu = x,$$

$$u(\epsilon x) = (u\epsilon)x = ux, \quad \epsilon x = x.$$

Hence ϵ is a principal unit. In such an algebra any number x for which $\Delta(x)\neq 0$ has a unique inverse. For, if x^{-1} is the unique number determined by $xx^{-1}=\epsilon$, then $x^{-1}x=\epsilon$, since

$$x(x^{-1}x - \epsilon) = \epsilon x - x\epsilon = 0.$$

Hence $x'x=\epsilon$ implies $x'=x^{-1}$, as shown by multiplying by x^{-1} on the right, so that $\Delta'(x)\neq 0$ (§ 6).

Thus $\Delta(x)\neq 0$ implies $\Delta'(x)\neq 0$ and conversely.

8. Transformation of units.

Consider n numbers

$$E_i = \sum_{j=1}^{n} c_{ij} e_j \qquad (i = 1, \ldots, n) \qquad (28)$$

in which the c's are numbers of a field F such that

$$|c_{ij}| \neq 0 \qquad (i, j = 1, \ldots, n).$$

We may solve the n equations and obtain

$$e_i = \sum_{j=1}^{n} t_{ij} E_j, \quad |t_{ij}| \neq 0 \quad (i = 1, \ldots, n) \quad (29),$$

where the t's are numbers of F. By means of (28) and (20) we can

* G. Scheffers, *Leipzig Berichte*, vol. 41 (1889), p. 293. Stated for commutative algebras by Weierstrass, *Göttingen Nach.*, 1884, p. 412.

express E_iE_j as a linear function of the e's, and hence by (29) as a linear function of the E's:

$$E_iE_j = \sum_{k=1}^{n} \Gamma_{ijk}E_k \quad (i, j = 1, \ldots, n) \qquad (30).$$

For any number x of the algebra,

$$x \equiv \sum_{i=1}^{n} x_ie_i = \sum_{j=1}^{n} X_jE_j, \quad X_j = \sum_{i=1}^{n} t_{ij}x_i \qquad (31).$$

Hence x can be expressed in one way and, in view of the linear independence of $E_1, \ldots, E_n$ with respect to F, but one way as ΣX_jE_j, where the X's are numbers of F. Taking $E_1, \ldots, E_n$ as new units, we obtain a linear algebra over F with the constants of multiplication Γ_{ijk}, which is called the *transform* of the initial algebra by the transformation of units (28) or (29). The two algebras are called *equivalent* under linear transformation of units in F.

9. Any number of a linear algebra is a root of an equation. Any $n+1$ numbers of a linear algebra with n units over a field F are linearly dependent with respect to F. For, if the first n of the $n+1$ numbers are linearly independent with respect to F, the $(n+1)$th number can be expressed as a linear function of them with coefficients in F (§ 8).

Assuming here that multiplication is associative, we may denote the product of i factors A by A^i, where A is any number of the algebra. Since $A, A^2, \ldots, A^{n+1}$ are linearly dependent, A is a root of an equation of degree $\leqq n+1$ with coefficients in F.

If also the algebra has a modulus ϵ, then $\epsilon, A, \ldots, A^n$ are linearly dependent and A is a root of an equation of degree $\leqq n$ with coefficients in F.

For example, in the case of the linear associative algebra of four units in § 4, $\epsilon = e_{11} + e_{22}$ is a principal unit, and the general number m, given by (16_1), is a root of

$$m^2 - (a+d)\,m + (ad - bc)\,\epsilon = 0.$$

10. Polynomials in a single number. An algebraic identity

$$f(x)\,g(x) \equiv p(x),$$

where the functions $f(x)$, etc., are polynomials in an ordinary complex variable with ordinary complex coefficients, without terms free of x, implies that the same relation holds when x is any number of a linear associative algebra. Indeed, the term involving x^k in $f(x)\,g(x)$ is obtained by multiplying the term in x^i of $f(x)$ by the term in x^{k-i} of $g(x)$ and summing the products for $i = 1, \ldots, k-1$. But the associative law gives $x^i x^{k-i} = x^k$.

The argument holds also for functions with terms free of the variable x provided the algebra has a principal unit ϵ and this is multiplied into those terms of the corresponding relation in the hyper-complex number x.

Since the associative law implies $x^r x^s = x^s x^r$, two polynomials in a hyper-complex number x are commutative.

11. Algebra of real quaternions; its unique place among algebras. We shall determine all linear associative algebras over the field of real numbers such that a product is zero only when one factor is zero. This determination is of decided intrinsic interest and yields an important result on real simple algebras (end of § 56).

If x is a given number $\neq 0$, $xx' = 0$ implies $x' = 0$; similarly, $x'x = 0$ implies $x' = 0$. Hence (§ 6) neither $\Delta(x)$ nor $\Delta'(x)$ is zero when $x \neq 0$. Thus (end of § 7) the algebra contains a principal unit, which we shall denote by 1. If every number is a real multiple of 1, the algebra is the field of all real numbers. Excluding this case, we may take the units to be $1, e_1, \ldots, e_{n-1}$, where $n > 1$. Then (§ 9) any number A of the algebra is a root of an equation $p(x) = 0$ of degree $\leqq n$ with real coefficients. By the fundamental theorem of algebra, $p(x)$ equals a product $f_1(x) f_2(x) \ldots$ of linear or quadratic factors with real coefficients. Then (§ 10), $f_1(A) f_2(A) \ldots = 0$. Thus one factor is zero. Hence any number of the algebra is a root of a quadratic equation with real coefficients.

If $e^2 + 2re + s = 0$, then $(e+r)^2 = r^2 - s$. Hence after adding a real constant to each e_i, we may assume that the square of each new unit e_i is a real number. If e_1^2 is a real number $\geqq 0$,

$$0 = e_1^2 - r^2 = (e_1 - r)(e_1 + r), \quad e_1 = \pm r,$$

whereas e_1 and 1 are linearly independent. Thus $e_1^2 = -t^2$, where t is a real number $\neq 0$. Set $E_1 = e_1/t$. Then $E_1^2 = -1$. If $n = 2$, the new units are $1, E_1 = i$, and the algebra is the system of ordinary complex numbers. Next, let $n > 2$. Then we may take the units to be $1, I, J, \ldots$, where*

$$I^2 = -1, \quad J^2 = -1, \ldots \qquad (32).$$

* Although $I^2 = J^2$, it does not follow that $(I-J)(I+J) = 0$, $I = \pm J$.

Since $I \pm J$ is a root of a quadratic equation,

$$(I+J)^2 \equiv -2+IJ+JI=r(I+J)+r_1,$$
$$(I-J)^2 \equiv -2-IJ-JI=s(I-J)+s_1,$$

where r, r_1, s, s_1 are real. Adding, we get

$$(r+s)I+(r-s)J+r_1+s_1+4=0.$$

But 1, I, J are linearly independent with respect to the field of reals. Hence $r=s=0$. Thus

$$IJ+JI=2c \qquad (c \text{ real}) \qquad (33).$$

The product IJ is linearly independent of 1, I, J with respect to the field of reals. For, if

$$IJ=r+sI+tJ,$$

where r, s, t are real, then

$$(I-t)\left(J+\frac{r+st}{t^2+1}I+\frac{rt-s}{t^2+1}\right)=0,$$

whereas neither factor is zero. Hence we may take

$$IJ=K \qquad (34)$$

as the fourth unit. In view of this choice of K, we do not know that $K^2=-1$, as in (32). But, by (32)—(34),

$$K^2=I(JI)J=I(2c-IJ)J=2cK-1,$$
$$(K-c)^2=c^2-1<0 \qquad (35).$$

For, if $c^2 \geqq 1$, $K-c$, and hence K, would be real.

We make the real transformation of units

$$i=I, \quad j=\frac{J+cI}{\sqrt{1-c^2}}, \quad k=\frac{K-c}{\sqrt{1-c^2}}.$$

Then $ij=k$, $ji=-k$. By (35), $k^2=-1$. By (32) and (33), $i^2=-1$, $j^2=-1$. Then the associative law gives

$$ik=i(ij)=-j, \quad ki=(ij)i=-ik=j,$$
$$kj=(ij)j=-i, \quad jk=j(ij)=-kj=i.$$

The resulting algebra, over the field of reals, with the four units 1, i, j, k and the multiplication table

$$\left.\begin{array}{llll} i^2=j^2=k^2=-1, & ij=k, & ji=-k, & \\ jk=i, & kj=-i, & ki=j, & ik=-j \end{array}\right\} \qquad (36),$$

is called the algebra Q of *real quaternions**.

* W. R. Hamilton, *Trans. Irish Acad.*, vol. 21 (1848), p. 199 [1843]; *Lectures*

Finally, suppose that our algebra contains a number λ not in this sub-algebra Q. Since λ satisfies a quadratic equation with real coefficients, we may derive as at the outset a number l, not in Q, such that $l^2 = -1$. By the proof leading to (33),

$$il + li = c_1, \quad jl + lj = c_2, \quad kl + lk = c_3,$$

where the c's are real constants. Then

$$lk = (li)j = (c_1 - il)j = c_1 j - i(c_2 - jl) = c_1 j - c_2 i + kl,$$

$$2kl = c_3 + c_2 i - c_1 j.$$

Multiply each term of the latter by k on the left. Thus

$$-2l = c_3 k + c_2 j + c_1 i.$$

But l was not in Q. We therefore have the

THEOREM*. *The only linear associative algebras over the field of reals, in which a product is zero only when one factor is zero, are the field of reals, the field of ordinary complex numbers, and the algebra of real quaternions.*

12. Simplest algebraic properties of real quaternions. To show that multiplication is associative, it suffices to verify (3_2) when each x is chosen from i, j, k. Instead of treating all 27 cases, it suffices, in view of the symmetry of (36), to treat the sets iii, iij, iji, jii, ijk. Now $(ij)k = -1 = i(jk)$, etc.

The *conjugate* q' of a quaternion q is defined by

$$q = x + yi + zj + wk, \quad q' = x - yi - zj - wk.$$

Their product is called the norm of q:

$$N(q) = qq' = q'q = x^2 + y^2 + z^2 + w^2.$$

If $q \neq 0$, then $N(q) \neq 0$ and the inverse of q is

$$q^{-1} = \frac{1}{N(q)} q'.$$

Thus, if $q \neq 0$, $qQ = q_1$ has the unique solution $Q = q^{-1}q_1$, and $Qq = q_1$

on Quaternions, 1853; *Elements of Quaternions*, 1866, etc. To assist the memory, note that if i, j, k be read in cyclic order (so that k is followed by i), the product of any one by the next is the next following one.

* Frobenius, *Crelle*, vol. 84 (1878), p. 59; C. S. Peirce, *Amer. Jour. Math.*, vol. 4 (1881), p. 225; E. Cartan, *Ann. Fac. sc. Toulouse*, vol. 12 (1898), B, p. 82 (see last foot-note in § 56); F. X. Grissemann, *Monatshefte Math. Phys.*, vol. 11 (1900), pp. 132—147 (the last a slight modification of the proof by Frobenius).

The remarkably simple proof in the text is due to the writer and was outlined by him in *Trans. Amer. Math. Soc.*, vol. 15 (1914), p. 39.

has the unique solution $Q=q_1q^{-1}$. In particular, a product of two quaternions is zero only when one factor is zero.

Since each kind of division by q is always possible and unique if and only if $N(q)\neq 0$, it is not surprising that a computation gives

$$\Delta(q)=\Delta'(q)=\{N(q)\}^2.$$

The conjugate of q' is q. The conjugate of qq_1 is $q_1'q'$. In their product we may move the real number q_1q_1' to the front of q. Hence the norm of a product of any two quaternions equals the product of their norms. This proves Euler's theorem that the product of two sums of four squares can be expressed as a sum of four squares. For applications, see end of § 52.

The general quaternion q is a root of

$$q^2-2xq+x^2+y^2+z^2+w^2=0.$$

A quaternion integer* is a quaternion

$$x(1+i+j+k)/2+yi+zj+wk,$$

in which $x, \ldots, w$ are ordinary integers. For them there is a greatest common divisor process, unique factorization into primes (apart from factors ± 1, $\pm i$, ..., which divide unity), etc., as in the arithmetic of ordinary integers.

13. Equivalence of the complex quaternion and matric algebras. We consider quaternions whose coordinates are ordinary complex numbers $a+b\sqrt{-1}$ and the complex matric algebra with four units (§ 4). From the units $e_{11}, \ldots$ of the matric algebra, we derive the quaternion units as follows:

$$1=e_{11}+e_{22},\quad i=\sqrt{-1}\,(e_{22}-e_{11}),\quad j=e_{12}-e_{21},\quad k=-\sqrt{-1}\,(e_{12}+e_{21}).$$

These satisfy the quaternion relations (36). Conversely,

$$e_{11}=\frac{1+\sqrt{-1}\,i}{2},\quad e_{22}=\frac{1-\sqrt{-1}\,i}{2},$$

$$e_{12}=\frac{j+\sqrt{-1}\,k}{2},\quad e_{21}=\frac{-j+\sqrt{-1}\,k}{2}.$$

The two complex algebras are therefore† equivalent under linear

* A. Hurwitz, *Göttingen Nachrichten*, 1896, p. 313.

† In § 45 of Cayley's paper (cited in § 3 above), he noted that relations (36) between the quaternion units can be satisfied by four matrices of order 2. B. Peirce, *Amer. Jour. Math.*, vol. 4 (1881), p. 132 (read before the Nat. Acad. Sc., 1870),

transformation of units, but the real quaternion and real matric sub-algebras are not.

14. Cayley's eight unit generalization of real quaternions. Cayley* gave the real algebra with the units 1, e_1, ..., e_7:

$$\left.\begin{array}{llll} e_i^2=-1, & e_ie_j=-e_je_i & (i, j=1, \ldots, 7;\ i\neq j) & \\ e_1e_2=e_3, & e_1e_4=e_5, & e_1e_6=e_7, & e_2e_5=e_7 \\ & e_2e_4=-e_6, & e_3e_4=e_7, & e_3e_5=e_6 \end{array}\right\} \quad (37),$$

together with 14 equations obtained from the last 7 by permuting each set of three subscripts cyclically, as $e_2e_3=e_1$, $e_3e_1=e_2$.

The norm $N(x)$ of a general number x is defined by

$$x=x_0+x_1e_1+\ldots+x_7e_7, \quad N(x)=x_0^2+x_1^2+\ldots+x_7^2.$$

While the associative law holds for three units chosen from one of the triples in (37), we have $e_ae_b \cdot e_c=-e_a \cdot e_be_c$ when the three e's are not in one of those triples. Since the associative law fails we may

knew that the complex quaternion algebra was equivalent to the linear algebra $e_{11}, e_{12}, e_{21}, e_{22}$ satisfying relations (14); he and C. S. Peirce knew the corresponding linear algebra (14) with n^2 units (pp. 217—18).

Sylvester, *Johns Hopkins Univ. Circulars*, vol. I (1882), p. 241; vol. III (1884), p. 7 (=*Math. Papers*, vol. III, p. 647; vol. IV, p. 122), noted that the matrices

$$\begin{pmatrix}1&0\\0&1\end{pmatrix}, \quad \begin{pmatrix}0&1\\-1&0\end{pmatrix}, \quad \begin{pmatrix}0&\theta\\ \theta&0\end{pmatrix}, \quad \begin{pmatrix}-\theta&0\\0&\theta\end{pmatrix} \quad [\theta=\sqrt{-1}]$$

"construed as complex numbers are a linear transformation of the ordinary quaternion system" [they satisfy (36) if taken as 1, i, k, j respectively, thus verifying Cayley's above remark], and gave the transformation in § 13. If ρ is a cube root of unity, the matrices

$$I=\begin{pmatrix}1&0&0\\0&1&0\\0&0&1\end{pmatrix}, \quad u=\begin{pmatrix}0&0&1\\ \rho&0&0\\0&\rho^2&0\end{pmatrix}, \quad v=\begin{pmatrix}0&0&1\\ \rho^2&0&0\\0&\rho&0\end{pmatrix}$$

satisfy the relations $u^3=v^3=I$, $vu=\rho uv$. He called the linear algebra with the nine units u^iv^j $(i, j=0, 1, 2)$ *nonions*. Sylvester gave another set of matrices u, v in *Compt. Rend. Paris*, vol. 97 (1883), p. 1336; vol. 98 (1884), pp. 273, 471 (=*Math. Papers*, vol. IV, pp. 118, 154). In *Amer. Jour. Math.*, vol. 6 (1884), p. 286 (=*Math. Papers*, vol. IV, p. 224), he called the matric algebra of order p, *p^2-ions*. Clifford had earlier given the name *quadrate algebra*.

* *Phil. Mag. London*, ser. 3, vol. 26 (1845), p. 210 (=*Coll. Math. Papers*, vol. I, p. 127). In his A_4, 87 should read 47. I have changed the sign of his last unit to obtain one (given by $\epsilon_7=-1$) of the two algebras later considered in more detail by Cayley, *Amer. Jour. Math.*, vol. 4 (1881), pp. 293—6 (=*Coll. Math. Papers*, vol. XI, pp. 368—371). His algebra with $\epsilon_7=+1$ may be obtained from that with $\epsilon_7=-1$ by changing the signs of e_2, ..., e_7. Hence the various Cayley algebras are equivalent to (37). An equivalent algebra was discovered independently by J. J. Graves before 1844, *Trans. Irish Acad.*, vol. 21 (1848), p. 338.

not prove by the method used for quaternions (end of § 12) that the norm of a product equals the product of the norms of the factors. To secure this property, Cayley made a long analysis which led finally to the choice of signs given in the relations (37). I have given a simple proof of this property and at the same time proved the remarkable theorem that right- and left-hand division, except by zero, are always possible and unique in this algebra, a fact overlooked by Cayley and first stated in a recent paper of mine*. I shall here give more elegant proofs based upon a representation of the 8 unit algebra as a quasi-binary algebra with real quaternion coordinates. Set $e = e_4$. Then the general number is $q + Qe$, where q, Q and r, R below are real quaternions in the units 1, e_1, e_2, e_3. It can be verified† that relations (37) imply

$$(q + Qe)(r + Re) = qr - R'Q + (Rq + Qr')\, e \qquad (38),$$

where r' is the quaternion conjugate to r.

Taking $r = q'$, $R = -Q$, we have

$$(q + Qe)(q' - Qe) = qq' + QQ' = N(q + Qe).$$

The norm of the product (38) is $tt' + TT'$, where

$$t = qr - R'Q, \quad T = Rq + Qr' \qquad (39),$$

and hence equals

$$(qq' + QQ')(rr' + RR') = N(q + Qe) \,.\, N(r + Re),$$

increased by $\alpha - \beta$, where

$$\alpha = RqrQ' + Qr'q'R', \quad \beta = qrQ'R + R'Qr'q'.$$

But the conjugate of the first term of α is the second term. Hence α is a real number. Thus $\alpha = R'\alpha R \div RR'$, and this is at once seen to equal β. Hence the norm of a product is the product of the norms.

Left-hand division except by zero is always possible and unique. For, if r, R, t, T be given, we can solve (39) for q, Q. To this end

* *Trans. Amer. Math. Soc.*, vol. 13 (1912), p. 72. Every number can be expressed as a linear function of e_2, e_4, e_6, with coefficients linear in e_1. If $B = r + se_1$, where r and s are real, set $\bar{B} = r - se_1$. Let also C be linear in e_1. Then

$$e_j B = \bar{B} e_j, \quad (Be_j)(Ce_j) = (B\bar{C})\, e_j^2, \quad (Be_j)(Ce_k) = (\bar{B}C)\,(e_j e_k),$$

for $j, k = 2, 4, 6$; $j \neq k$. Hence the 8 unit algebra can be exhibited as a quasi-quaternion algebra with the units 1, e_2, e_4, e_6 and coordinates linear in e_1.

† The reader may take (38) as the definition of the algebra.

multiply the second equation (39) by r on the right and replace qr by its value from the first; we get

$$(rr' + RR') Q = Tr - Rt.$$

Multiply the first by r' on the right and eliminate Qr'. Thus

$$(rr' + RR') q = tr' + R'T.$$

Similarly, right-hand division except by zero is always possible and unique. By the relations in the first line of (37), the general number x is a root of

$$x^2 - 2x_0 x + N(x) = 0.$$

The above theorem on the norm of a product and the corresponding one for norms of quaternions (end of § 12) and for norms of ordinary complex numbers lead to identities

$$(x_1^2 + \dots + x_n^2)(y_1^2 + \dots + y_n^2) = z_1^2 + \dots + z_n^2$$

in the x's and y's, where $z_1, \dots, z_n$ are bilinear functions of the x's and the y's, for the cases $n = 2, 4, 8$. That there is such an identity only in these three cases was proved by A. Hurwitz*, using the theory of matrices. Many earlier writers had published unsatisfactory proofs†. All writers have overlooked the initial paper by C. F. Degen‡, who gave the identity for $n = 8$, and a method which he supposed would succeed for $n = 16$.

15. Characteristic determinants and equations. Let $e_1, \dots, e_n$ be the units of a linear associative algebra over a field F and having a principal unit ϵ. Let $x = \Sigma x_i e_i$ be the general number of the algebra. By (20),

$$xe_j = \sum_{k=1}^{n} y_{jk} e_k, \quad y_{jk} = \sum_{i=1}^{n} x_i \gamma_{ijk} \qquad (j = 1, \dots, n).$$

For $j = 1, \dots, n$ in turn, the first equation gives

$$(y_{11} - x) e_1 + y_{12} e_2 + \dots + y_{1n} e_n = 0,$$

$$\dots\dots\dots\dots\dots\dots\dots\dots$$

$$y_{n1} e_1 + y_{n2} e_2 + \dots + (y_{nn} - x) e_n = 0.$$

Let ω be an arbitrary number of F. When x is replaced by ω, the determinant of the coefficients of the e's becomes $\delta(\omega)$ in (40). Let $C_1(\omega), \dots, C_n(\omega)$ be the cofactors of the elements of the jth column.

* *Göttingen Nachrichten*, 1898, p. 309.

† A partial list is in *Encyclopédie Sc. Math.*, vol. I, 1, pp. 368, 467.

‡ *Mém. Acad. Sc. St. Pétersbourg*, vol. 8, années 1817—18 (1822), p. 207. There is a misprint in the sign of Rt.

Multiply the linear equations by $C_1(x)\epsilon, \ldots, C_n(x)\epsilon$ on the left. Then (§ 10)

$$\delta(x) e_j = 0 \qquad (j = 1, \ldots, n).$$

Multiply by ϵ_j and sum, where $\Sigma \epsilon_j e_j = \epsilon$ Thus $\delta(x)\epsilon = 0$. Starting with $e_j x$, we obtain the second part of

THEOREM 1. *In a linear associative algebra with a principal unit, the general number x is a root of the right-hand and left-hand characteristic equations $\delta(\omega) = 0$, $\delta'(\omega) = 0$, respectively, where*

$$\delta(\omega) = \left| \sum_{i=1}^{n} \gamma_{ijk} x_i - \delta_{jk}\omega \right|, \quad \delta'(\omega) = \left| \sum_{i=1}^{n} \gamma_{jik} x_i - \delta_{jk}\omega \right| \qquad (j, k = 1, \ldots, n) \quad (40).$$

Here δ_{jk} is Kronecker's symbol (§ 7). These determinants are called the *right- and left-hand characteristic determinants*, respectively. The same definitions are made for any linear algebra. Thus the characteristic determinants are derived from the determinants $\Delta(x)$ and $\Delta'(x)$ of § 6 by subtracting ω from each element of the main diagonals.

For any linear algebra with a principal unit $\epsilon = \Sigma \epsilon_i e_i$, there is a deeper relation between these pairs of determinants. Then

$$\sum_{i=1}^{n} (x_i - \omega\epsilon_i)\gamma_{ijk} = \sum_{i=1}^{n} x_i \gamma_{ijk} - \omega\delta_{jk},$$

$$\sum_{i=1}^{n} (x_i - \omega\epsilon_i)\gamma_{jik} = \sum_{i=1}^{n} x_i \gamma_{jik} - \omega\delta_{jk},$$

by (26). Hence

$$\Delta(x - \omega\epsilon) \equiv \delta(\omega), \quad \Delta'(x - \omega\epsilon) \equiv \delta'(\omega) \qquad (41).$$

As multiplication may not be associative, we shall write

$$x^1 = x, \quad x^{i+1} = x^i x, \quad {}^1x = x, \quad {}^{i+1}x = x({}^i x) \qquad (i = 1, 2, \ldots),$$

and, in case there is a modulus ϵ, $x^0 = {}^0x = \epsilon$. If $\Sigma t_i x^i = 0$, x is called a *right-hand root* of $\Sigma t_i \omega^i = 0$. If $\Sigma t_i ({}^i x) = 0$, x is called a *left-hand root* of $\Sigma t_i \omega^i = 0$. We shall prove

THEOREM 2. *In any linear algebra with a principal unit, the general number is a left-hand root of the right-hand characteristic equation and a right-hand root of the left-hand characteristic equation.*

Without loss of generality (§ 16), we may assume that the principal unit is $e_1 = 1$. By (41), $\delta(\omega)$ and $\delta'(\omega)$ are derived from $\Delta(x)$ and $\Delta'(x)$ by replacing x_1 by $x_1 - \omega$. Set

$$\delta(\omega) = \Delta(x - \omega) = \sum_{i=0}^{n} r_i \omega^i.$$

Denote by $x_1', \ldots, x_n'$ the cofactors of the elements of the first row in

$$\Delta(x) = \begin{vmatrix} \Sigma x_i \gamma_{i11} \ldots \Sigma x_i \gamma_{in1} \\ \ldots\ldots\ldots\ldots\ldots\ldots \\ \Sigma x_i \gamma_{i1n} \ldots \Sigma x_i \gamma_{inn} \end{vmatrix} \tag{42}$$

Multiply x_j' by the jth element $\Sigma x_i \gamma_{ijk}$ of the kth row and sum for $j = 1, \ldots, n$. We obtain the coefficient y_k of e_k in $y = xx'$, where $x' = \Sigma x_j' e_j$. But $y_1 = \Delta(x)$, $y_k = 0$ $(k > 1)$. Hence $xx' = \Delta(x)$. When x_1 is replaced by $x_1 - \omega$, let x' become

$$f = \sum_{i=0}^{n-1} f_i \omega^i,$$

where the f_i are numbers of the algebra. Thus

$$(x - \omega) f = \Delta(x - \omega) = \sum_{i=0}^{n} r_i \omega^i.$$

Expanding the first product and equating the coefficients of like powers of ω, we get

$$xf_0 = r_0, \quad xf_1 - f_0 = r_1, \ldots, \quad xf_{n-1} - f_{n-2} = r_{n-1}, \quad -f_{n-1} = r_n.$$

Multiply the second equation by x on the left, the third by x twice on the left, the fourth by x three times, etc. Adding, we get

$$\Sigma r_i ({}^i x) = 0.$$

The proof of the second part of the theorem † is similar.

A final generalization, not hitherto published, is

THEOREM 3. *In an arbitrary linear algebra, the general number is a left-hand root of* $\omega\delta(\omega) = 0$ *and a right-hand root of* $\omega\delta'(\omega) = 0$.

Let $e_1, \ldots, e_n$ be the units of the given algebra A, which is not assumed to have a principal unit, nor to be associative. Consider the algebra A^* with the units $e_0, e_1, \ldots, e_n$, where

$$e_0^2 = e_0, \quad e_0 e_i = e_i e_0 = e_i \qquad (i = 1, \ldots, n).$$

† This theorem was stated and proved by Dickson, *Trans. Amer. Math. Soc.*, vol. 13 (1912), p. 60. For the case in which multiplication is associative, the proof becomes essentially the simpler one of the two proofs by Frobenius, *Sitzungsber. Akad. Berlin*, 1896, p. 601, when that proof is suitably translated from the terminology of bilinear forms into that of hyper-complex numbers. In this associative case, the theorem can be expressed as one on matrices and was first stated in this form by Cayley and verified for $n=2$ and $n=3$, *Phil. Trans. London*, vol. 148 (1858), p. 24 (= *Coll. Math. Papers*, vol. II, p. 475). For matrix m in (7), the theorem is $m^2 - (a+d)m + (ad - bc)\begin{pmatrix} 1 & 0 \\ 0 & 1 \end{pmatrix} \equiv 0$. For references to many other proofs, see *Encyc. Sc. Math.*, vol. I, 1, p. 418. The earliest occurrence of the theorem was, for $n=3$, in Hamilton's *Lectures on Quaternions*, 1853, pp. 566—7.

Thus e_0 is a principal unit of A^*. Set

$$x = \sum_{i=1}^{n} x_i e_i, \quad x^* = x_0 e_0 + x, \quad x' = \sum_{i=1}^{n} x_i' e_i.$$

Then x and x^* are the general numbers of A and A^*. Since

$$(x_0' e_0 + x') x^* = x_0' x_0 e_0 + \sum_{i=1}^{n} (x_0' x_i + x_i' x_0) e_i + x' x,$$

the left-hand determinant of x^* is

$$\Delta'_{n+1}(x^*) = \begin{vmatrix} x_0 & 0 & \cdots & 0 \\ x_1 & x_0 + \Sigma x_j \gamma_{1j1} & \cdots & \Sigma x_j \gamma_{nj1} \\ \cdots & \cdots & \cdots & \cdots \\ x_n & \Sigma x_j \gamma_{1jn} & \cdots & x_0 + \Sigma x_j \gamma_{njn} \end{vmatrix},$$

with x_0 occurring only in the main diagonal. By (41), the left-hand characteristic determinant $D(\omega)$ of x^* is therefore

$$\Delta'_{n+1}(x^* - \omega e_0) = (x_0 - \omega) \sum_{i=0}^{n} l_i (\omega - x_0)^i,$$

where $\Sigma l_i \omega^i$ is the expansion of the left-hand characteristic determinant $\delta'(\omega)$ of x. By the second theorem above, x^* is a right-hand root of $D(\omega) = 0$. Set $x_0 = 0$. Hence x is a right-hand root of $\Sigma l_i \omega^{i+1} = 0$. The first part of the theorem is proved similarly.

16. Determinants of x unaltered by linear transformation of units. Introduce new units $E_1, \ldots, E_n$ as in § 8; let x become $X = \Sigma X_j E_j$. We are to prove that $\Delta(x)$, given by (23), equals

$$D(X) = \left| \sum_{i=1}^{n} X_i \Gamma_{ijk} \right| \qquad (j, k = 1, \ldots, n).$$

If x and y are given numbers of the initial algebra, $xx' = y$ has a unique solution x' if and only if $\Delta(x) \neq 0$ (§ 6). Similarly, $XX' = Y$ has a unique solution X' if and only if $D(X) \neq 0$. But $x = X$, etc. Hence $\Delta(x) = 0$ implies $D(X) = 0$. If $\Delta(x)$ is identically zero, then $D(X)$ is, and the theorem $\Delta(x) = D(X)$ is true. Henceforth, let $\Delta(x)$ be not identically zero, so that there exists a modulus ϵ (§ 7). Since every set of values of the x_i, γ_{ijk} for which $\Delta(x) = 0$ is a set of solutions of $D(X) = 0$, it will follow from a well-known theorem on polynomials that $\Delta(x)$ is a factor of $D(X)$ as soon as $\Delta(x)$ is proved to be an irreducible function of its arguments x_i, γ_{ijk}. Since each element of the determinant $D(X)$ equals a function linear and homogeneous in $x_1, \ldots, x_n$, and linear and homogeneous in the γ's, the quotient of $D(X)$ by $\Delta(x)$ depends only upon the coefficients t_{ij} of the transformation of units. The quotient is unity, since the latter is its value for $x = \epsilon$ by (27).

It remains only to prove that $\Delta(x)$ is irreducible. If the function $\Delta(x)$, homogeneous in the γ's and homogeneous in $x_1, \ldots, x_n$, is the product of two factors, each factor is homogeneous in the γ's and in the x's. But the coefficient of x_1^n is $|\gamma_{1jk}|$, a determinant whose n^2 elements are arbitrary and hence an irreducible function of these n^2 elements γ_{1jk}. Hence one factor of Δ is free of the γ's. Taking $\gamma_{ijk} = 0$ $(i \neq j)$, we have

$$\Delta(x) = |x_j \gamma_{jjk}| = x_1 \ldots x_n |\gamma_{jjk}| \qquad (j, k = 1, \ldots, n).$$

Hence for general values of the γ's, the factor free of the γ's is $x_1 \ldots x_n$. But the coefficient of x_1^n in Δ was seen to be not identically zero. A like argument holds for $\Delta'(x)$. Hence $\Delta(x)$ and $\Delta'(x)$ are unaltered under linear transformation of the units. Since the modulus ϵ is unaltered, we have*, by (41), the

THEOREM. *The determinants and characteristic determinants of the general number of any linear algebra remain unaltered under any linear transformation of the units.*

17. Invariants and covariants of linear algebras. Consider the linear algebra with n units whose constants of multiplication† γ_{ijk} are undetermined numbers of a field F. Let C be a polynomial in these γ's and the coordinates x_i of the general number x of the algebra over F. If, under every linear transformation of units (§ 8),

$$C(X_i;\ \Gamma_{ijk}) = f \cdot C(x_i;\ \gamma_{ijk}),$$

where f is a function only of the coefficients t_{ij} of the transformation, C is called a *covariant* of the algebra. In particular, C is an absolute covariant if $f = 1$, and an invariant if C involves only the γ's.

The characteristic determinants $\delta(\omega)$ and $\delta'(\omega)$ are absolute covariants of the general linear algebra with n units (§ 16).

For example, consider the algebra with the units ϵ, e, where ϵ is a principal unit, and $e^2 = \gamma e + c\epsilon$. For $x = x_1\epsilon + x_2 e$,

$$\Delta(x) = \Delta'(x) = \begin{vmatrix} x_1 & cx_2 \\ x_2 & x_1 + \gamma x_2 \end{vmatrix},$$

$$\delta(\omega) = \delta'(\omega) = \omega^2 - l\omega + \Delta(x), \quad l \equiv 2x_1 + \gamma x_2.$$

* The invariance of $\delta(\omega)$ when $\Delta(x) \equiv 0$ follows by continuity.

† If there be a modulus, we take it as the unit e_1. Then

$$\gamma_{1jj} = \gamma_{j1j} = 1, \quad \gamma_{1jk} = \gamma_{j1k} = 0 \quad (j \neq k).$$

The remaining γ's are to be left arbitrary. See the example.

Then l and $\Delta(x)$ are absolute covariants. To give a direct proof, introduce the new units ϵ, $E = re + s\epsilon$, where $r \neq 0$. Then

$$x = X_1\epsilon + X_2E, \quad X_1 = x_1 - \frac{sx_2}{r}, \quad X_2 = \frac{x_2}{r},$$

$$E^2 = \Gamma E + C\epsilon, \quad \Gamma = 2s + r\gamma, \quad C = r^2c - rs\gamma - s^2,$$

$$\begin{vmatrix} X_1 & CX_2 \\ X_2 & X_1 + \Gamma X_2 \end{vmatrix} = \Delta(x), \quad 2X_1 + \Gamma X_2 = l.$$

The discriminant of $\Delta(x)$ gives an invariant of the algebra :

$$\Gamma^2 + 4C = r^2(\gamma^2 + 4c).$$

18. Binary linear algebras with a principal unit. In the last example, take $r = 1$, $s = -\gamma/2$. Then $\Gamma = 0$. Let therefore $\gamma = 0$ in the initial algebra. In order that the transformed algebra shall have $\Gamma = 0$, the transformation must be $E = re$ (i.e. have $s = 0$). Then $C = r^2c$. Hence the binary algebra

$$\epsilon^2 = \epsilon, \quad \epsilon e = e\epsilon = e, \quad e^2 = c\epsilon \tag{43}$$

and the similar one with the parameter C, are equivalent if and only if c and C are both zero, or both are not zero and their ratio is the square of a number of the field F.

If F is the field of all complex numbers, the two types of non-equivalent binary algebras are (43) with $c = 0$, $c = 1$.

If F is the field of reals, the three types are (43) with $c = 0$, $c = 1$, $c = -1$, the last being the field $F(i)$ of complex numbers.

Cayley* gave the seven types of non-equivalent binary associative algebras over the field of complex numbers, the presence of a principal unit not being assumed. Miss O. C. Hazlett† recently obtained these seven types from the triple algebras having a modulus (§ 20) and characterized them by covariants.

19. Rank and rank equation of a linear algebra. Let $e_1, \ldots, e_n$ be the units of an algebra over a field F and let the coordinates $x_1, \ldots, x_n$ of $x = \Sigma x_i e_i$ be undetermined numbers of F.

First, let the algebra be associative. By § 15, x is a root of $\delta(\omega) = 0$ or $\omega\delta(\omega) = 0$, according as the algebra has or has not a principal unit. Thus x^n or x^{n+1} is a linear combination of lower powers of x (also by § 9). The *rank* r of the algebra is the least positive integer such that x^r is a linear combination of lower powers

* *Proc. London Math. Soc.*, vol. 15 (1883—4), p. 185 (= *Coll. Math. Papers*, vol. XII, pp. 60, 105).

† *Annals of Mathematics*, vol. 16 (1914), p. 1.

of x whose coefficients are rational functions of $x_1, \ldots, x_n$ with coefficients in F. The equation $R(x) = 0$ itself is called the *rank equation* of the algebra*. If there were two such equations, the difference would give an equation of lower degree.

Now $R(\omega)$ is a divisor of $\delta(\omega)$ or $\omega\delta(\omega)$. For, if not, the division by $R(\omega)$ would lead to a remainder $\rho(\omega)$, of degree $< r$, which vanishes for $\omega = x$, contrary to the definition of r (cf. § 10).

Similarly, $R(\omega)$ divides $\delta'(\omega)$ or $\omega\delta'(\omega)$.

It follows that the coefficients of $R(\omega)$ are integral rational functions of $x_1, \ldots, x_n$. For, if not, some set of finite values of the x's would give an infinite root of a characteristic equation.

Let $\omega_1, \ldots, \omega_r$ be the ordinary complex roots of $R(\omega) = 0$. Then, if the algebra has a modulus ϵ,

$$R(x) \equiv (x - \omega_1\epsilon) \ldots (x - \omega_r\epsilon) = 0.$$

The factors are commutative (§ 10). Thus $f = x - \omega_i\epsilon$ is a number $\neq 0$ such that $fg = 0$ for some $g \neq 0$. Such a number f is called a *nil factor*†. Then $\Delta(f) = 0$, and since $gf = 0$, $\Delta'(f) = 0$ (§ 6). Hence, by (41), $\omega_1, \ldots, \omega_r$ are roots of the two characteristic equations. Conversely, any root ω of one of the latter, say $\delta(\omega) = 0$, is one of the set $\omega_1, \ldots, \omega_r$. For, if $w = x - \omega\epsilon$, then $\Delta(w) = 0$ by (41) and we can find a number $y \neq 0$ such that $wy = 0$. Now

$$v_i \equiv x - \omega_i\epsilon = w + \mu_i\epsilon, \quad \mu_i \equiv \omega - \omega_i,$$

$$0 = v_1 v_2 \ldots v_r = (\quad) w + \mu_1\mu_2 \ldots \mu_r\epsilon.$$

Multiply this by y on the right. Thus $0 = \mu_1 \ldots \mu_r y$. Hence one of the μ's is zero. We thus have (Scheffers, *l.c.*)

THEOREM 1. *For a linear associative algebra having a principal unit, the distinct roots of the rank equation $R(\omega) = 0$ are identical with the distinct roots of either characteristic equation.*

For example, the results in § 12 show that each characteristic determinant of the quaternion $x + yi + zj + wk$ is

$$\{(x-\omega)^2 + y^2 + z^2 + w^2\}^2 = \{R(\omega)\}^2.$$

* Th. Molien, *Math. Annalen*, vol. 41 (1893), p. 113. G. Scheffers, *ibid.*, vol. 39 (1891), p. 293, called it the characteristic equation of the system, and r the "Grad." It is often called the identical equation of the algebra.

† B. Peirce, *Amer. Jour. Math.*, vol. 4 (1881), p. 104; Weierstrass, *Gött. Nach.*, 1884, p. 395.

Under a transformation of units, let the rank equation

$$R(\omega;\ x_i,\ \gamma_{ijk}) = 0,$$

satisfied by $\omega = x$, be transformed into $\rho(\omega;\ X_i,\ \Gamma_{ijk}) = 0$, so that $\rho = 0$ for $\omega = X$. But $R(\omega;\ X_i,\ \Gamma_{ijk}) = 0$ for $\omega = X$. Unless the last equation is identical with $\rho = 0$, we obtain by subtraction an equation of degree $< r$ satisfied by $\omega = X$. This equation is the transform of an equation of degree $< r$ satisfied by $\omega = x$, contrary to the definition of r. Hence (Miss Hazlett, *l.c.*):

THEOREM 2. *The rank equation of a linear associative algebra is unaltered by every linear transformation of units.*

In the sense of § 17, the rank r is an invariant of the general associative algebra with n units, whereas the rank function $R(\omega)$ is not a covariant.

For non-associative linear algebras, there is an equation $\rho(\omega) = 0$ of lowest degree having x as a right-hand root (§ 15), called the right-hand rank equation. Then* $\rho(\omega)$ divides $\delta'(\omega)$ or $\omega\delta'(\omega)$, according as there is or is not a modulus. A similar definition and property hold for the left-hand rank equation. These equations are unaltered by every linear transformation of units.

For example, in the commutative, but not associative, linear algebra with the units $\epsilon, e_1, \ldots, e_5$, of which ϵ is a principal unit, and

$$e_1^2 = e_2,\quad e_1e_2 = e_3,\quad e_1e_3 = -e_2,\quad e_1e_4 = e_5,\quad e_1e_5 = e_3^2 = -e_4,$$
$$e_2^2 = e_4,\quad e_2e_3 = -e_5,\quad e_2e_4 = e_2e_5 = e_3e_4 = e_3e_5 = e_4^2 = e_4e_5 = e_5^2 = 0,$$

$w = x_1e_1 + \ldots + x_5e_5$ is a root of

$$w \,.\, w^3 + x_1^2w^2 = 0,$$

but not of a cubic or quadratic equation. The general number $x = x_0\epsilon + w$ is therefore a root of

$$x \,.\, x^3 - 4x_0x^3 + (x_1^2 + 6x_0^2)\,x^2 - 2\,(x_0x_1^2 + 2x_0^3)\,x + (x_0^2x_1^2 + x_0^4)\,\epsilon = 0.$$

20. Complex ternary linear associative algebras with a modulus†. Consider a linear associative algebra, over the field of all complex numbers, with three units and the modulus ϵ. Its rank r is 3 or 2.

(I) First, let $r = 3$. There exists a number α such that

$$R(\alpha) \equiv (\alpha - \lambda_1\epsilon)(\alpha - \lambda_2\epsilon)(\alpha - \lambda_3\epsilon) = 0,$$

and such that α is not a root of a quadratic equation.

* *Trans. Amer. Math. Soc.*, vol. 13 (1912), p. 62, Cor. II.

† E. Study, *Göttingen Nach.*, 1889, pp. 243—7. Many computations there made or left to the reader are avoided in the exposition in this tract. Study gave no argument leading to (A), (B) in case II.

(I$_1$) If the roots $\lambda_1, \lambda_2, \lambda_3$ are distinct, set

$$e_1 = \frac{(a-\lambda_2\epsilon)(a-\lambda_3\epsilon)}{(\lambda_1-\lambda_2)(\lambda_1-\lambda_3)}, \quad e_2 = \frac{(a-\lambda_1\epsilon)(a-\lambda_3\epsilon)}{(\lambda_2-\lambda_1)(\lambda_2-\lambda_3)}, \quad e_3 = \frac{(a-\lambda_1\epsilon)(a-\lambda_2\epsilon)}{(\lambda_3-\lambda_1)(\lambda_3-\lambda_2)}.$$

For a an arbitrary number, the sum of these e's will be proved to equal ϵ. If we replace a by $\lambda_i\epsilon$ $(i=1, 2, 3)$, we see that

$$e_1+e_2+e_3=\epsilon \qquad (44)$$

is satisfied. But an algebraic quadratic equation is an identity if it is satisfied by three values. Thus (44) holds. By the associative law, powers of a and hence the $a-\lambda_i\epsilon$ are commutative. Thus, if $i \neq j$, e_ie_j has the factor R and is zero. Multiplying (44) on the right by e_i, we get $e_i^2=e_i$. Hence

$$e_ie_j=0, \quad e_i^2=e_i \qquad (i=1, 2, 3;\ i\neq j) \qquad (45).$$

Finally, the e's are linearly independent. For, by multiplying $\Sigma a_ie_i=0$ by e_j on the right, we get $a_je_j=0$, $a_j=0$, since $e_j\neq 0$. The resulting algebra (45) is evidently associative, and has

$$R=\delta=\delta'=(x_1-\omega)(x_2-\omega)(x_3-\omega).$$

(I$_2$) If $\lambda_1=\lambda_2\neq\lambda_3$, set $l=\lambda_1-\lambda_3$ and

$$e_1=-(a-\lambda_3\epsilon)(a-2\lambda_1\epsilon+\lambda_3\epsilon)/l^2,$$
$$e_2=c(a-\lambda_1\epsilon)(a-\lambda_3\epsilon), \quad e_3=(a-\lambda_1\epsilon)^2/l^2,$$

where c is any constant $\neq 0$. Each product in the first line of

$$\left.\begin{aligned} e_1e_3=e_3e_1=e_2e_3=e_3e_2=e_2^2=0 \\ e_1^2=e_1, \quad e_1e_2=e_2e_1=e_2, \quad e_3^2=e_3 \end{aligned}\right\} \qquad (46)$$

has the factor R and is zero. By addition, $e_1+e_3=\epsilon$. Multiplying this on the left by e_1, e_2, e_3 in turn and on the right by e_2, we get the relations in the second line of (46). The units are linearly independent since the products of $\Sigma a_ie_i=0$ on the left by e_2 and e_3 in turn give $a_1=a_3=0$. The relations (46) for this associative algebra are unaltered if e_2 is replaced by te_2. Here

$$R=\delta=\delta'=(x_1-\omega)^2(x_3-\omega).$$

(I$_3$) If $\lambda_1=\lambda_2=\lambda_3$, set $e_1=\epsilon$, $e_2=a-\lambda_1\epsilon$, $e_3=(a-\lambda_1\epsilon)^2$. Then

$$e_1e_i=e_ie_1=e_i, \quad e_2e_3=e_3e_2=e_3^2=0, \quad e_2^2=e_3 \qquad (47),$$
$$R=\delta=\delta'=(x_1-\omega)^3.$$

(II) Next, let $r=2$. Take $e_1=\epsilon$, e_2, e_3 as the units. Since $r=2$, $\xi=ye_2+ze_3$ is a root of a quadratic $\xi^2+2L\xi+Q_1\epsilon=0$, where L is a linear and Q_1 a quadratic function of y and z with fixed coefficients depending upon the γ_{ijk}. Thus $(\xi - L\epsilon)^2 = Q\epsilon$, where Q is a quadratic function of y, z. If Q is not identically zero it can be transformed linearly into yz or z^2, neither of which is zero for $(y, z) = (1, 1)$ or $(1, 2)$. Hence we can find two linearly independent sets (y, z) for which $Q \neq 0$, and therefore three linearly independent units ϵ, $\xi_2 - L_2\epsilon$, $\xi_3 - L_3\epsilon$, such that the squares of the last two are $Q_2\epsilon$, $Q_3\epsilon$, respectively, where $Q_2Q_3 \neq 0$. Take $E_i = (\xi_i - L_i\epsilon)/Q_i^{\frac{1}{2}}$. Then

$$E_2^2 = \epsilon, \quad E_3^2 = \epsilon \qquad \text{(A)}.$$

But if Q, and hence the square of $ye_2 + ze_3 - L\epsilon$, is identically zero, where L is a certain linear function of y, z, we have only to subtract constant multiples of ϵ from e_2, e_3 to obtain new units for which

$$e_2^2 = 0, \quad e_3^2 = 0 \qquad \text{(B)}.$$

(II_1) Consider case (A). Set

$$E_2E_3 = \alpha\epsilon + \beta E_2 + \gamma E_3.$$

Then

$$E_2(E_2E_3) = (\beta + \alpha\gamma)\,\epsilon + (\alpha + \beta\gamma)\,E_2 + \gamma^2 E_3 = E_2^2E_3 = E_3,$$

$$(E_2E_3)\,E_3 = (\gamma + \alpha\beta)\,\epsilon + \beta^2 E_2 + (\alpha + \beta\gamma)\,E_3 = E_2E_3^2 = E_2,$$

$$\beta^2 = \gamma^2 = 1, \quad \alpha + \beta\gamma = \beta + \alpha\gamma = \gamma + \alpha\beta = 0.$$

When the sign of E_2 is changed, that of γ is changed. Hence we may set $\gamma = +1$. Changing if necessary the sign of E_3, we may also set $\beta = +1$. Hence

$$E_2E_3 = -\epsilon + E_2 + E_3.$$

Set $e_2 = E_2$, $e_3 = E_2 + E_3$. Then

$$e_2^2 = \epsilon, \quad e_2e_3 = e_3, \quad e_3^2 - e_3e_2 = e_3,$$

$$e_3e_2 = r\epsilon + se_2 + te_3.$$

Equating the two values of $e_2e_3e_2$ and the two values of $e_3e_2e_2$, we get $r = s$, $t^2 = 1$, $s(1 + t) = 0$, respectively. For $t = 1$, $e_3e_2 = e_3$, $e_3^2 = 2e_3$, and ϵ, $f \equiv e_2 + e_3$, $f^2 = \epsilon + 4e_3$ are linearly independent, whereas the rank is 2. Hence $t = -1$,

$$e_3e_2 = s\epsilon + se_2 - e_3, \quad e_3^2 = s\epsilon + se_2,$$

$$(e_3e_2)\,e_3 = 2se_3 - e_3^2 = e_3(e_2e_3) = e_3^2, \quad e_3^2 = se_3,$$

so that $s = 0$. We have therefore the algebra

$$e_1e_i = e_ie_1 = e_i, \quad e_2^2 = e_1, \quad e_2e_3 = e_3, \quad e_3e_2 = -e_3, \quad e_3^2 = 0 \qquad (48),$$

$$\delta = lR, \quad \delta' = l'R, \quad R = ll', \quad l \equiv x_1 + x_2 - \omega, \quad l' = x_1 - x_2 - \omega.$$

Here, and in (49), R may be derived from δ and δ' as in § 19.

(II$_2$) Finally, consider case (B). Let e, f denote e_2, e_3 or e_3, e_2. Set $ef = a\epsilon + be + cf$. Then

$$0 = e^2 f = e(ef) = ac\epsilon + (a + bc)e + c^2 f, \quad c = a = 0,$$
$$0 = ef^2 = (ef)f = (be)f = b^2 e, \quad b = 0,$$
$$e_1 e_i = e_i e_1 = e_i, \quad e_2^2 = e_2 e_3 = e_3 e_2 = e_3^2 = 0 \qquad (49),$$
$$\delta = \delta' = (x_1 - \omega)^3, \quad R = (x_1 - \omega)^2.$$

No two of the five resulting linear associative algebras (45)—(49) *with a principal unit are equivalent. They are characterized by the invariant* r *and the covariant* $\delta(\omega)$.

The corresponding problem for 4, 5, 6 units has been treated*.

21. Reducible linear associative algebras with a modulus. A linear associative algebra A with n units over a field F and with a modulus ϵ is called *reducible*† with respect to F if it contains $p + q = n$ numbers $e_1, \ldots, e_p$; $E_1, \ldots, E_q$, linearly independent with respect to F, such that

$$e_i E_j = 0, \quad E_j e_i = 0 \qquad (i = 1, \ldots, p;\ j = 1, \ldots, q) \quad (50).$$

Any number of A is a linear combination of the e's and E's. Let the modulus be $\epsilon = e + E$, where e is a linear function of $e_1, \ldots, e_p$, and E of $E_1, \ldots, E_q$. If x is any linear function of $e_1, \ldots, e_p$ with coefficients in F, then $x = x\epsilon = xe$, since $xE = 0$, and $x = \epsilon x = ex$. Similarly, if X is any linear function of $E_1, \ldots, E_q$, then $EX = XE = X$. Next, $x_1 x_2 = x + X$, where x_1 and x_2 are any linear functions of $e_1, \ldots, e_p$. Multiply by E on the right. We get $0 = X$. Hence‡ the product of any two x's is an x. Similarly, the product of any two X's is an X. Hence the numbers x form a sub-algebra $s = (e_1, \ldots, e_p)$ with a modulus e, and the numbers X form a sub-algebra $S = (E_1, \ldots, E_q)$ with a modulus E. The algebra A is said to be *decomposable* into s and S, and is called their *direct*§ *sum*, $A = s + S = S + s$.

* *Encyc. Sc. Math.*, vol. I, 1, pp. 401—3. For the irreducible algebras with six units, see G. Voghera, *Denkschr. Ak. Wiss.*, Wien, vol. 84 (1908).

† G. Scheffers, *Math. Annalen*, vol. 39 (1891), p. 317; vol. 41 (1893), p. 601. In *Amer. Jour. Math.*, vol. 4 (1881), p. 100, B. Peirce defined a mixed (impure) algebra A to be one each of whose numbers is the sum of a number of a sub-algebra s and a number of a sub-algebra S such that the products eE and Ee of any numbers e of s and E of S are numbers common to s and S. In case zero is the only common number, the mixed algebra is reducible in the sense of the text.

‡ Proved less simply by S. Epsteen, *Trans. Amer. Math. Soc.*, vol. 5 (1904), p. 105.

§ A mixed algebra is not always the direct sum of s and S.

Conversely, from any two linear (associative) algebras $(e_1, \dots, e_p)$ and $(E_1, \dots, E_q)$ over F with moduli e and E, we obtain a linear (associative) algebra $(e_1, \dots, E_q)$ over F with the modulus $e+E$ by postulating relations (50) and regarding $e_1, \dots, E$ to be linearly independent with respect to F.

Scheffers gave the following criterion for reducibility:

A linear associative algebra A with a modulus ϵ is reducible if and only if it contains a number $e \neq \epsilon$ such that $e^2 = e$, $ex = xe$, for every number x of A.

That these conditions are necessary was proved above. That they are sufficient is proved by setting $E = \epsilon - e$ and showing that $A = s + S$, where s is composed of all products xe, and S of all products xE, x ranging over all numbers of A. We have

$$eE = e(\epsilon - e) = e - e = 0, \quad Ee = (\epsilon - e)e = 0,$$

$$xe \,.\, yE = xy \,.\, eE = 0, \quad yE \,.\, xe = y(Ee)x = 0,$$

since $ey = ye$ for any y in A. Also $A = s + S$ since

$$x = x\epsilon = x(e+E) = xe + xE.$$

Finally, if the xe are expressible linearly in terms of linearly independent numbers $e_1, \dots, e_p$ of the form xe, and the xE in terms of $E_1, \dots, E_q$, then $e_1, \dots, E_q$ are linearly independent. For, if $xe + yE = 0$, the product by e on the right gives $xe = 0$.

A component s of a reducible algebra A may be reducible or irreducible. Hence a reducible algebra may be decomposed into irreducible algebras. For example, algebra (45) is the sum of three irreducible algebras (e_1), (e_2), (e_3), and, by (44), its modulus is the sum of the moduli e_1, e_2, e_3 of the sub-algebras.

Concerning the uniqueness of decomposition, see end of § 61.

22. Direct product of two algebras. Let $s = (e_1, \dots, e_p)$, $S = (E_1, \dots, E_q)$ be two linear algebras over a field F, so that

$$e_i e_j = \sum_{k=1}^{p} \gamma_{ijk} e_k, \quad E_i E_j = \sum_{k=1}^{q} \Gamma_{ijk} E_k.$$

Proceeding* in a formal manner, regard $\epsilon_{ik} = e_i E_k = E_k e_i$, for $i = 1, \dots, p$; $k = 1, \dots, q$, as pq linearly independent units of an algebra P with

* W. K. Clifford, *Amer. Journ. Math.*, vol. 1 (1878), p. 350 (= *Coll. Math. Papers*, 1882, p. 266). If $\omega^2 = 1$ and ω is commutative with every real quaternion q and Q, he called $q + \omega Q$ a *biquaternion* (not Hamilton's biquaternion, a quaternion with ordinary complex coefficients).

coordinates ranging over F, and set

$$\epsilon_{ik}\epsilon_{jl} = e_i e_j \,.\, E_k E_l \equiv \sum_{g=1}^{p} \sum_{h=1}^{q} \gamma_{ijg} \Gamma_{klh} \epsilon_{gh}.$$

Call P the *direct* product* of s and S, and write $P = sS = Ss$.

We may also obtain P by interpreting the coordinates $x_1, \ldots, x_p$ in $x = x_1 e_1 + \ldots + x_p e_p$ to be general numbers

$$x_i = X_{i1} E_1 + \ldots + X_{iq} E_q \qquad (X\text{'s in } F)$$

of S. Thus $x = \Sigma X_{ik} e_i E_k$ $(i = 1, \ldots, p;\ k = 1, \ldots, q)$. Or we may regard P to be an algebra of q units over the algebra s.

For example, from the real quaternion algebra $(1, i, j, k)$ and the real algebra $(1, \sqrt{-1})$, we obtain the complex quaternion algebra.

If s and S are associative algebras, $P = sS$ is associative. If s and S have moduli e and E, P has the modulus eE.

For this definition of multiplication of algebras and that of addition in § 21, the commutative, associative and distributive laws hold.

THEOREM 1†. *If A is a linear associative algebra having the quaternion algebra Q as a sub-algebra and having the same modulus* 1 *as Q, then $A = QC$, where C is a sub-algebra, with the modulus* 1, *of A such that every number of C is commutative with every number of Q.*

It is just as easy to prove Wedderburn's generalization :

THEOREM 2. *If A is a linear associative algebra with the modulus ϵ and the sub-algebra Q with n^2 units $e^a f^b$ $(a, b = 1, \ldots, n)$:*

$$e^n = f^n = \epsilon, \quad fe = \rho e f \qquad (51),$$

where ρ is a primitive nth root of unity, then $A = QC$, where C is a sub-algebra, with the same modulus ϵ, of A such that every number of C is commutative with every number of Q.

The inverse of $E_{a,b} \equiv e^a f^b$ is $\rho^{ab} E_{-a,-b}$ since (by induction)

$$f^b e^a = \rho^{ab} e^a f^b \qquad (51').$$

If X is any number of A, the n^2 numbers

$$N_{c,d} = \sum_{a,b=1}^{n} E^{-1}_{a,b} E_{c,d} X E_{a,b} \qquad (c, d = 1, \ldots, n)$$

* To distinguish it from the non-commutative "product" $\Sigma z_{ij} e_i E_j$ of any two sets $\Sigma x_i e_i$ and $\Sigma y_j E_j$ of numbers of an algebra.

† G. Scheffers, *Math. Ann.*, vol. 39 (1891), pp. 364—374. In place of this proof occupying ten pages, we shall give the short proof due to J. H. M. Wedderburn, *Proc. Roy. Soc. Edinb.*, vol. 26, I (1905—6), p. 48, of a more general theorem.

are commutative with every number of Q. Indeed,

$$E_{r,s}N_{c,d} = \sum_{a,b=1}^{n} \rho^{ab} e^r f^s e^{-a} f^{-b} E_{c,d} X E_{a,b} = \sum_{a,b} \rho^{ab-as} e^{r-a} f^{s-b} E_{c,d} X E_{a,b}$$

$$= \sum_{\alpha,\beta=1}^{n} \rho^{r\beta+\alpha\beta} e^{-\alpha} f^{-\beta} E_{c,d} X E_{r+\alpha,\, s+\beta} = N_{c,d} E_{r,s},$$

in which we have replaced the summation indices a, b by $r+\alpha$, $s+\beta$, respectively. We find at once that

$$E^{-1}{}_{c,d} E^{-1}{}_{a,b} E_{c,d} = \rho^{ad-bc} E^{-1}{}_{a,b}.$$

Summing for $c, d = 1, \ldots, n$, we get zero unless a and b are multiples of n, i.e. unless $E_{a,b} = \epsilon$, and then we get $n^2\epsilon$. Hence

$$\sum_{c,d} E^{-1}{}_{c,d} N_{c,d} = n^2 X.$$

The set C of all numbers $\lambda, \mu, \ldots$ of A which are commutative with every X of A contains ϵ, $\lambda+\mu$, $\lambda\mu$, and hence is a sub-algebra with the modulus ϵ. We saw that C contains every $N_{c,d}$ and that X is expressible as a sum of products of numbers $E^{-1}{}_{c,d}$ of Q by numbers $N_{c,d}$ of C. Hence every X of A is in the direct product QC. The product is direct since

$$\sum_{a,b=1}^{n} E^{-1}{}_{a,b}\gamma_{a,b} = 0 \qquad (\gamma\text{'s in } C)$$

implies that each $\gamma = 0$. Since γ is commutative with the E's,

$$0 = \sum_{a,b} \sum_{c,d} E^{-1}{}_{c,d} E^{-1}{}_{a,b} E_{c,d} \gamma_{a,b} = n^2 \gamma_{nn}.$$

To prove that $\gamma_{\alpha\beta} = 0$, we multiply the given relation by $E_{\alpha,\beta}$ and use the resulting relation in which $\gamma_{\alpha\beta}$ is the coefficient of $\epsilon = E^{-1}{}_{n,n}$.

The complex algebra defined by (51) is equivalent to the complex matric algebra of § 4. Indeed, (51) are satisfied if

$$e = e_{12} + e_{23} + \ldots + e_{n1}, \quad f = e_{11} + \rho^{-1}e_{22} + \ldots + \rho^{-(n-1)}e_{nn}.$$

The algebra (51) is Sylvester's algebra of nonions if $n = 3$ (§ 13).

23. Units normalized relatively to a fixed number. Given a number $a = \Sigma a_i e_i$ of any algebra whose coordinates are *scalars*, i.e. ordinary complex numbers, we can find a number

$$y = \Sigma y_i e_i \neq 0$$

and a scalar ω such that

$$ay = \omega y \tag{52}$$

By (20) and the linear independence of the e's, necessary and sufficient conditions for (52) are

$$\sum_{i,j=1}^{n} \gamma_{ijk} a_i y_j - \omega y_k = 0 \qquad (k = 1, \ldots, n).$$

The determinant of the coefficients of $y_1, \ldots, y_n$ is $\delta(\omega)$, given by (40) with $x = a$. Hence $\delta(\omega) = 0$ is a necessary and sufficient condition for the existence of a number $y \neq 0$ satisfying (52).

For the general number x let $\delta(\omega) = 0$ define ω as an h-valued function of $x_1, \ldots, x_n$, as in the theory of algebraic functions. Let a be a particular number x for which there are h distinct roots $\omega_1, \ldots, \omega_h$ of multiplicities $m_1, \ldots, m_h$.

If y_1 is a second solution of (52), then $cy + c_1 y_1$ is a solution, where c and c_1 are any scalars. Hence all solutions y are linear functions of certain t linearly independent solutions, which we take as the first t of our new units $e_1, \ldots$. Then $ae_j = \omega_1 e_j$ $(j \leqq t)$ give

$$\sum_{i=1}^{n} a_i \gamma_{ijk} = \omega_1 \delta_{jk} \qquad (k = 1, \ldots, n;\ j = 1, \ldots, t) \qquad (53),$$

where, as usual, $\delta_{jj} = 1$, $\delta_{jk} = 0$ $(k \neq j)$. Then (§ 16)

$$\delta(\omega) = \left| \sum_{i=1}^{n} a_i \gamma_{ijk} - \omega \delta_{jk} \right| = \left| \begin{array}{cccc:c} \omega_1 - \omega & 0 & \ldots & 0 & \\ 0 & \omega_1 - \omega & \ldots & 0 & \\ \ldots & \ldots & \ldots & \ldots & \\ 0 & 0 & \ldots & \omega_1 - \omega & \ldots \\ 0 & 0 & \ldots & 0 & M \\ \ldots & \ldots & \ldots & \ldots & \end{array} \right|,$$

in the first t columns of which all elements are zero except the diagonal elements $\omega_1 - \omega$, while M denotes the matrix of the elements in the last $n - t$ rows and last $n - t$ columns. Hence

$$\delta(\omega) = (\omega_1 - \omega)^t \, |M| = 0$$

has ω_1 as a root of multiplicity m_1, where $m_1 \geqq t$.

If $m_1 = t$, we proceed no further with ω_1 (see the example in § 24). Next, let $m_1 > t$. Then there exist numbers $z = \Sigma z_j e_j \neq 0$ such that

$$az = \omega_1 z + \sum_{k=1}^{t} c_k e_k \qquad (54):$$

$$\sum_{i,j,k=1}^{n} a_i z_j \gamma_{ijk} e_k = \omega_1 \sum_{k=1}^{n} z_k e_k + \sum_{k=1}^{t} c_k e_k.$$

The coefficients of e_k $(k = 1, \ldots, t)$ give equations which serve to determine the c_k. Consider the coefficient of e_k $(k > t)$; in it the coefficient of $z_j (j \leqq t)$ is zero by (53), since $k>j$. We thus have the conditions

$$\sum_{j=t+1}^{n} \sum_{i=1}^{n} a_i \gamma_{ijk} z_j - \omega_1 z_k = 0 \qquad (k = t+1, \ldots, n).$$

The matrix of the coefficients of the z's is the above M if $\omega = \omega_1$. But $|M| = 0$. Hence there exist $t + t'$ $(t' \geqq 1)$ linearly independent solutions z of (54), including the solutions $e_1, \ldots, e_t$ of (52). Taking these as the first $t + t'$ new e's, we find as before that $\delta(\omega)$ has the factor $(\omega_1 - \omega)^{t+t'}$. If $m_1 = t + t'$, we proceed no further with ω_1. But if $m_1 > t + t'$, there exist solutions $w \neq 0$ of

$$aw = \omega_1 w + \sum_{k=1}^{t+t'} d_k e_k.$$

Ultimately we reach m_1 linearly independent numbers

$$\alpha_1, \alpha_1', \ldots, \alpha_1^{(m_1-1)}$$

such that

$$a\alpha_1^{(i)} = \omega_1 \alpha_1^{(i)} + \text{lin. func. of } \alpha_1, \ldots, \alpha_1^{(i-1)} \qquad (55).$$

Similarly, if ω_j is a root of multiplicity m_j, there exist m_j linearly independent numbers $\alpha_j, \alpha_j', \ldots$, such that*

$$a\alpha_j^{(i)} = \omega_j \alpha_j^{(i)} + \text{lin. func. of } \alpha_j, \ldots, \alpha_j^{(i-1)} \qquad (55').$$

These $n = m_1 + \ldots + m_h$ numbers $\alpha_j^{(i)}$ are linearly independent. First, if $\alpha_2 = l$, where $l = c_0 \alpha_1 + \ldots + c_k \alpha_1^{(k)}$, $c_k \neq 0$, then

$$\omega_2 \alpha_2 = a\alpha_2 = \omega_1 l + l', \quad 0 = (\omega_1 - \omega_2)\, l + l',$$

where l' is a linear function of $\alpha_1, \ldots, \alpha_1^{(k-1)}$. But $\alpha_1^{(k)}$ is not a linear function of those α's. Next, if $\alpha_2' = c\alpha_2 + l$, multiplication by a on the left gives

$$\omega_2 \alpha_2' + g\alpha_2 = c\omega_2 \alpha_2 + \omega_1 l + l'.$$

Eliminating α_2', we get

$$g\alpha_2 = (\omega_1 - \omega_2)\, l + l',$$

which was proved impossible if $g \neq 0$ or $g = 0$. The general step of the proof follows similarly by induction. Hence the $\alpha_j^{(i)}$ may be introduced as n units normalized relatively to a.

By the proof leading to (55), if ax differs from $\omega_1 x$ by a linear function of $\alpha_1, \alpha_1', \ldots$, then x itself is such a linear function.

* For the case of associative algebras having a modulus, this was stated without proof by E. Cartan, *Ann. Fac. Sc. Toulouse*, vol. 12 (1898), memoir B, p. 17.

Occasionally we shall use the result similar to our first one : there is a number $y \neq 0$ and a scalar ω such that

$$yx = \omega y \qquad (56)$$

if and only if $\delta'(\omega) = 0$.

24. Example. The algebra with the units $e_1, \ldots, e_5$, such that

$$e_1 e_5 = e_1, \quad e_2 e_3 = e_1, \quad e_2 e_5 = e_2, \quad e_3 e_5 = e_3,$$
$$e_4 e_1 = e_1, \quad e_4 e_2 = e_2, \quad e_4^2 = e_4, \quad e_5 e_3 = e_3, \quad e_5^2 = e_5,$$

while the remaining $e_i e_j$ are zero, is associative and has the modulus $\epsilon = e_4 + e_5$. The conditions for $xy = \omega y$ are

$$(x_4 - \omega) y_1 + x_2 y_3 + x_1 y_5 = 0, \quad (x_4 - \omega) y_2 + x_2 y_5 = 0,$$
$$(x_5 - \omega) y_3 + x_3 y_5 = 0, \quad (x_4 - \omega) y_4 = 0, \quad (x_5 - \omega) y_5 = 0.$$

The determinant of the coefficients of $y_1, \ldots, y_5$ is evidently

$$\delta(\omega) \equiv (x_4 - \omega)^3 (x_5 - \omega)^2.$$

For $x = a = e_5$, $x_5 = 1$, $x_i = 0$ $(i < 5)$, and $\delta(\omega) = 0$ has the roots 0, 1.

If $\omega = 0$, the five conditions reduce to $y_3 = y_5 = 0$. Since y_1, y_2, y_4 are arbitrary, we may take e_1, e_2, e_4 as the linearly independent solutions y of $ay = 0$. Next, if $\omega = 1$, the conditions reduce to $y_1 = y_2 = y_4 = 0$, and we may take e_3, e_5 as the linearly independent solutions y of $ay = y$. Hence e_1, e_2, e_4, e_3, e_5 are units normalized relatively to $a = e$.

PART II

REVISION OF CARTAN'S GENERAL THEORY OF COMPLEX LINEAR ASSOCIATIVE ALGEBRAS WITH A MODULUS

25. Units having a character. If x is any number, § 23 gives

$$a(a_1 x) = (aa_1)\, x = (\omega_1 a_1)\, x = \omega_1 (a_1 x).$$

Hence (end of § 23), $a_1 x$ is a linear function of $a_1, a_1', \ldots$. Next,

$$a(a_1' x) = (\omega_1 a_1' + \lambda a_1)\, x = \omega_1 (a_1' x) + \text{lin. func. of } a_1, a_1', \ldots.$$

Hence $a_1' x$ is a linear function of $a_1, a_1', \ldots$. By induction, we see similarly that any $a_j^{(i)} x$ is a linear function of $a_j, a_j', \ldots$.

If ϵ is the modulus, $\epsilon = \epsilon_1 + \ldots + \epsilon_h$, where ϵ_j is a linear function of the units $a_j, a_j', \ldots$. By the preceding result, $\epsilon_j a_1^{(i)}$ is a linear function of $a_j, a_j', \ldots$. Hence

$$a_1^{(i)} = \epsilon a_1^{(i)} = \sum_{j=1}^{h} \epsilon_j a_1^{(i)}$$

gives

$$\epsilon_1 a_1^{(i)} = a_1^{(i)}, \quad \epsilon_j a_1^{(i)} = 0 \qquad (j \neq 1) \qquad (57).$$

Similarly, we have

$$\epsilon_2 a_2^{(i)} = a_2^{(i)}, \quad \epsilon_j a_2^{(i)} = 0 \qquad (j \neq 2), \ldots \qquad (58).$$

It follows at once that

$$\epsilon_j^2 = \epsilon_j, \quad \epsilon_j \epsilon_k = 0 \qquad (j \neq k) \qquad (59).$$

We shall call $\epsilon_1, \ldots, \epsilon_h$ *partial moduli* of the algebra.

Since $\epsilon_1 \epsilon_1 = \epsilon_1$, not every $a_1^{(i)} \epsilon_1$ is zero. After a suitable rearrangement of $a_1, a_1', \ldots$, we may assume that

$$a_1 \epsilon_1, \ a_1' \epsilon_1, \ \ldots, \ a_1^{(p-1)} \epsilon_1 \qquad (p \geqq 1) \qquad (60)$$

are linearly independent, while each $a_1^{(k)} \epsilon_1$ ($k \geqq p$) is linearly dependent on the numbers (60):

$$a_1^{(k)} \epsilon_1 = c_0^{(k)} a_1 \epsilon_1 + \ldots + c^{(k)}_{p-1} a_1^{(p-1)} \epsilon_1 \qquad (k = p, \ldots, m_1 - 1).$$

In place of these units $\alpha_1^{(k)}$ we introduce

$$\bar{\alpha}_1^{(k)} = \alpha_1^{(k)} - c_0^{(k)}\alpha_1 - \dots - c^{(k)}_{p-1}\alpha_1^{(p-1)} \qquad (k = p, \dots, m_1 - 1).$$

Now $\bar{\alpha}_1^{(k)}\epsilon_1 = 0$. Dropping the bars from these α's, we have m_1 linearly independent units $\alpha_1^{(i)}$ $(i = 0, 1, \dots, m_1 - 1)$ such that

$$\alpha_1^{(k)}\epsilon_1 = 0 \qquad (k = p, \dots, m_1 - 1) \qquad (61),$$

while the numbers (60) are linearly independent.

The property that any $\alpha_1^{(i)}x$ is a linear function of $\alpha_1, \alpha_1', \dots$ evidently holds true also for the present $\alpha_1^{(i)}$. Set

$$\alpha_1\epsilon_1 = k_0\alpha_1 + \dots + k_{m_1-1}\alpha_1^{(m_1-1)}.$$

Since $(\alpha_1\epsilon_1)\epsilon_1 = \alpha_1\epsilon_1$, we have, by (61),

$$\alpha_1\epsilon_1 = k_0\alpha_1\epsilon_1 + \dots + k_{p-1}\alpha_1^{(p-1)}\epsilon_1.$$

By the linear independence of the numbers (60),

$$k_0 = 1, \quad k_1 = 0, \dots, \quad k_{p-1} = 0,$$

$$\alpha_1\epsilon_1 = \alpha_1 + k_p\alpha_1^{(p)} + \dots + k_{m_1-1}\alpha_1^{(m_1-1)}.$$

Similarly,

$$\alpha_1'\epsilon_1 = \alpha_1' + k_p'\alpha_1^{(p)} + \dots, \qquad \alpha_1^{(p-1)}\epsilon_1 = \alpha_1^{(p-1)} + k_p^{(p-1)}\alpha_1^{(p)} + \dots.$$

The right members $\bar{\alpha}_1, \bar{\alpha}_1', \dots, \bar{\alpha}_1^{(p-1)}$ of these equations, together with $\alpha_1^{(p)}, \dots, \alpha_1^{(m_1-1)}$, evidently give m_1 linearly independent functions of the $\alpha_1^{(i)}$, and hence may be taken as new units. But, by (61),

$$\bar{\alpha}_1\epsilon_1 = \alpha_1\epsilon_1 = \bar{\alpha}_1, \quad \bar{\alpha}_1'\epsilon_1 = \bar{\alpha}_1', \dots, \quad \bar{\alpha}_1^{(p-1)}\epsilon_1 = \bar{\alpha}_1^{(p-1)}.$$

Dropping the bars from these α's, we have (61) and

$$\alpha_1\epsilon_1 = \alpha_1, \quad \alpha_1'\epsilon_1 = \alpha_1', \dots, \quad \alpha_1^{(p-1)}\epsilon_1 = \alpha_1^{(p-1)}.$$

Since $\epsilon_1\epsilon_i = 0$ by (59), if $i \neq 1$,

$$\alpha_1\epsilon_i = (\alpha_1\epsilon_1)\epsilon_i = \alpha_1(\epsilon_1\epsilon_i) = 0, \quad \alpha_1'\epsilon_i = 0, \dots, \quad \alpha_1^{(p-1)}\epsilon_i = 0 \qquad (i \neq 1).$$

Hence if η is any one of the numbers $\alpha_1, \alpha_1', \dots, \alpha_1^{(p-1)}$,

$$\eta\epsilon_1 = \eta, \quad \eta\epsilon_i = 0, \quad \epsilon_1\eta = \eta, \quad \epsilon_i\eta = 0 \qquad (i \neq 1) \qquad (62),$$

the last pair from (57). These hold also for $\eta = \epsilon_1$ by (59).

If $p = m_1$, the η's include all the $\alpha_1^{(j)}$. But if $p < m_1$, $\alpha_1^{(k)}\epsilon_2, \dots, \alpha_1^{(k)}\epsilon_h$ are not all zero, where k is any fixed integer $\geqq p$. For, if so, $\alpha_1^{(k)} = \alpha_1^{(k)}\epsilon = \alpha_1^{(k)}\epsilon_1$, contrary to (61). Suppose first that not all of the $\alpha_1^{(k)}\epsilon_2$ are zero. Then the argument beginning with (60) is repeated for $\alpha_1^{(p)}, \dots, \alpha_1^{(m_1-1)}$ in their relation to ϵ_2. Thus if η is any one of certain $q > 0$ numbers $\alpha_1^{(p)}, \dots, \alpha_1^{(p+q-1)}$,

$$\eta\epsilon_2 = \eta, \quad \eta\epsilon_i = 0 \ (i \neq 2), \quad \epsilon_1\eta = \eta, \quad \epsilon_i\eta = 0 \ (i \neq 1) \qquad (63),$$

the last pair from (57). We proceed similarly until the set

$$\alpha_1, \alpha_1', \ldots, \alpha_1^{(m_1-1)}$$

is exhausted.

An analogous distribution of the $\alpha_2^{(i)}$ into sets of η's may be made, and likewise for the $\alpha_h^{(i)}$. We may assume that ϵ_2 is one of the η's coming from the $\alpha_2^{(i)}$, etc.

Any η in (62) is said to have the character (1, 1); any η in (63) the character (1, 2). In general, η is of *character** (α, β) if

$$\epsilon_i\eta = 0 \ (i \neq \alpha), \quad \epsilon_\alpha\eta = \eta, \quad \eta\epsilon_j = 0 \ (j \neq \beta), \quad \eta\epsilon_\beta = \eta \qquad (64).$$

In particular, ϵ_i is of character (i, i).

THEOREM. *We can find n linearly independent units* $\epsilon_1, \ldots, \epsilon_h, \eta_1, \ldots, \eta_{n-h}$ *each having a definite character.*

For example, in the matric algebra of n^2 units e_{ij} in § 4, the $\epsilon_i = e_{ii}\,(i = 1, \ldots, n)$ are the partial moduli, and e_{ij} is of character (i, j) in view of (14).

Any number η is said to be of character (α, β) if and only if it satisfies relations (64). *The sum of two numbers of like character has that character; the sum of two numbers of unlike character has no character.* For, if η has the character (α, β) and η' the character (γ, δ), $(\eta + \eta')\,\epsilon_\beta = \eta + \eta'$ or η, according as $\beta = \delta$ or $\beta \neq \delta$; while

$$\epsilon_\alpha(\eta + \eta') = \eta + \eta' \text{ or } \eta,$$

according as $\gamma = \alpha$ or $\gamma \neq \alpha$.

26. Example. We shall find the character of each of the normalized units in the example of § 24. Here $h = 2$, $\epsilon_1 = e_4$, $\epsilon_2 = e_5$. Then

$$\epsilon_1^2 = \epsilon_1, \quad \epsilon_2^2 = \epsilon_2, \quad \epsilon_1\epsilon_2 = \epsilon_2\epsilon_1 = 0,$$
$$\epsilon_1 e_1 = e_1, \quad \epsilon_2 e_1 = 0, \quad e_1\epsilon_1 = 0, \quad e_1\epsilon_2 = e_1,$$
$$\epsilon_1 e_2 = e_2, \quad \epsilon_2 e_2 = 0, \quad e_2\epsilon_1 = 0, \quad e_2\epsilon_2 = e_2,$$
$$\epsilon_1 e_3 = 0, \quad \epsilon_2 e_3 = e_3, \quad e_3\epsilon_4 = 0, \quad e_3\epsilon_2 = e_3.$$

Hence ϵ_1 is of character (1, 1), ϵ_2 and e_3 are of character (2, 2), e_1 and e_2 of character (1, 2).

27. Theorem. *The product of a number η of character (α, β) by a number η' of character (γ, δ) is zero if $\beta \neq \gamma$, and is either zero or is a number of character (α, δ) if $\beta = \gamma$.*

* Introduced by G. Scheffers, *Math. Ann.*, vol. 39 (1891), p. 313, in connection with algebras without a quaternion sub-algebra. The present proof of the general theorem is an amplification of that by Cartan, *l. c.*, p. 19.

We have $\epsilon_a \eta = \eta \epsilon_\beta = \eta, \quad \epsilon_\gamma \eta' = \eta' \epsilon_\delta = \eta',$

$$\epsilon_a \eta\eta' = \eta\eta' = \eta\eta' \epsilon_\delta, \quad \eta\eta' = (\eta\epsilon_\beta)(\epsilon_\gamma \eta') = \eta(\epsilon_\beta \epsilon_\gamma)\eta'.$$

Hence, if $\beta \neq \gamma$, $\eta\eta' = 0$ since $\epsilon_\beta \epsilon_\gamma = 0$. If $\beta = \gamma$ and $\eta\eta' \neq 0$, then $\eta\eta'$ is of character (a, δ).

28. The sub-algebras Σ_i. The product of two numbers of character (i, i) is of character (i, i). In view also of the remark at the end of § 25, all of the numbers of character (i, i) of the initial algebra Σ form a linear associative sub-algebra Σ_i with the modulus ϵ_i.

The characteristic equation for Σ_1 has a single root and its multiplicity is the number m of units of Σ_1.*

For, if it had two or more distinct roots, Σ_1 would have at least two partial moduli ϵ_1', ϵ_1'' (§ 25). Set

$$a = x_1' \epsilon_1' + x_1'' \epsilon_1'' + x_2 \epsilon_2 + \dots + x_h \epsilon_h,$$

where the x's are arbitrary scalars. Then

$$a\epsilon_1' = x_1' \epsilon_1', \quad a\epsilon_1'' = x_1'' \epsilon_1'', \quad a\epsilon_2 = x_2 \epsilon_2, \ \dots, \ a\epsilon_h = x_h \epsilon_h.$$

Hence (§ 23) the characteristic equation of a for Σ has the distinct roots x_1', x_1'', x_2, ..., x_h, whereas the number is $\leqq h$.

29. Choose the units e_i of Σ_1 so that e_1 is the modulus ϵ_1. As just proved, the characteristic equation of $x = \Sigma x_i e_i$ is $(\omega - l)^m = 0$, where $l = \Sigma c_i x_i$. Taking $x = \epsilon_1$, we have $\omega = 1$, $l = c_1$, whence $c_1 = 1$. Take as new units ϵ_1, $\eta_i = e_i - c_i \epsilon_1$ $(i = 2, \dots, m)$. Then

$$x = x_1' \epsilon_1 + \Sigma x_i \eta_i,$$

where $x_1' = l$. For the algebra in the new units, the characteristic equation of x is $(\omega - x_1')^m = 0$. Hence the characteristic equation of any linear combination of $\eta_2, \dots, \eta_m$ has no root other than zero.

30. Nilpotent numbers. A number is called *nilpotent*† if some power of it is zero. Let $\eta^k = 0$, $\eta \neq 0$, and call ω any root of the characteristic equation of η. There exists (§ 23) a number $y \neq 0$ such that $\eta y = \omega y$. Multiply by η^{k-1} on the left. Thus

$$0 = \omega\eta^{k-1} y = \omega\eta^{k-2}(\omega y) = \omega^2 \eta^{k-3}(\omega y) = \dots = \omega^k y.$$

Hence $\omega = 0$. Conversely, if the characteristic equation of η has all its roots zero, a power of η vanishes by Theorem 1, § 15. Hence

* The prefix "right hand" will often be omitted.

† B. Peirce, *Amer. Journ. Math.*, vol. 4 (1881), p. 97. The name *Wurzel der Null* was used by Frobenius, *Sitzungsb. Ak. Berlin*, 1903, p. 635; *pseudo-nul* by Cartan, *l.c.*, p. 21.

a number is nilpotent if and only if every root of its characteristic equation is zero. Similarly, or by Theorem 1 of § 19, *every root of the left-hand characteristic equation of a nilpotent number is zero.*

31. Theorem. *As the units of Σ_1 we may take its modulus ϵ_1 and $m-1$ nilpotent numbers η such that any linear function of the η's is nilpotent.*

This follows from the result in § 29.

32. Normalized units. In view of the last theorem and that in § 25, we may take the n units of any algebra to be the partial moduli $\epsilon_1, \ldots, \epsilon_h$ and $n-h$ nilpotent numbers each of a definite character. Note that any number of character (α, β), $\alpha \neq \beta$, is nilpotent, since its square is zero.

33. Examples. In the algebra considered in §§ 24, 26, the sub-algebra Σ_2 of numbers of character (2, 2) has the two units ϵ_2 and e_3, of which ϵ_2 is the partial modulus. For $x=e_3$, the formula in § 24 gives $\delta(\omega)=\omega^5$. Thus e_3 is nilpotent in the main algebra Σ. We have $e_3{}^2=0$. Similarly, e_1 and e_2 are nilpotent. Hence the normalized units of Σ are ϵ_1, ϵ_2 and the nilpotent numbers e_1, e_2, e_3.

As an instructive example, the reader may show by means of the general theory that the algebra of complex quaternions is equivalent to the matric algebra of four units [take $a=i$ in § 23].

He may treat also the algebra $(e_1, \ldots, e_4)$ where

$$e_1e_4=e_1, \quad e_2e_3=e_2, \quad e_3e_1=e_1, \quad e_3{}^2=e_3, \quad e_4{}^2=e_4, \quad e_4e_2=e_2,$$

while all further products of two units are zero; the modulus is e_3+e_4.

34. Theorem. *The product of any nilpotent number η of character (i, i) by any number u of the same character is nilpotent.*

For concreteness, take $i=1$. Then the product ηu is of character (1, 1) and hence is in algebra Σ_1. By § 31,

$$\eta u = a\epsilon_1 + \eta' \qquad (a \text{ scalar}, \eta' \text{ nilpotent}).$$

Since $\eta\epsilon_1=\eta$, we have

$$\eta^2 u = a\eta + \eta\eta'.$$

Let m be the least positive integer for which $\eta^m u=0$. If $m=1$, the theorem is true. If $m \geqq 2$, multiply the preceding equation by η^{m-2} on the left. Thus

$$\eta^{m-1}\eta' = (-a)\,\eta^{m-1}, \quad \eta^{m-1} \neq 0.$$

Hence $a=0$ since η' is nilpotent (end of § 30). Thus $\eta u = \eta'$.

In the example in § 26, § 33, e_3 is nilpotent, e_3 and ϵ_2 are of character (2, 2), and $e_3\epsilon_2=e_3$ is nilpotent.

35. Separation of algebras into two categories. An algebra is of the first category if the determinant $\Delta(x)$ of its general number $\Sigma x_i e_i$ is the product of n linear homogeneous functions* of $x_1, \ldots, x_n$.

An algebra is of the second category if $\Delta(x)$ has a non-linear irreducible factor $f(x_1, \ldots, x_n)$.

A binary algebra is of the first category since a homogeneous function of x_1 and x_2 is a product of linear functions. The five types of ternary algebras (45)—(49) are of the first category, since for each $\Delta(x) \equiv \delta(0)$ is a product of linear factors.

The algebra of complex quaternions is of the second category, since (§ 12) $\Delta(q)$ is the square of the irreducible function $x^2+y^2+z^2+w^2$. We may also employ the equivalent matric algebra of § 4. By (16), (17),

$$\Delta(m) = \begin{vmatrix} a & 0 & b & 0 \\ 0 & a & 0 & b \\ c & 0 & d & 0 \\ 0 & c & 0 & d \end{vmatrix} = \begin{vmatrix} a & b & 0 & 0 \\ c & d & 0 & 0 \\ 0 & 0 & a & b \\ 0 & 0 & c & d \end{vmatrix} \qquad (65),$$

while a general determinant of order two is irreducible.

We shall later (§ 46) prove that an algebra is of the second category if and only if it has a quaternion sub-algebra. It will then follow that the above classification into algebras of the first and second categories (Cartan, *l. c.*, p. 24) coincides with that† of algebras into those without and with quaternion sub-algebras.

ALGEBRAS A_1 OF THE FIRST CATEGORY.

36. Theorem. *The sum of two nilpotent numbers of A_1 is nilpotent.*

The distinct roots of the characteristic equation of $x = \Sigma x_i e_i$ are by hypothesis of the form

$$\omega_i = a_{i1} x_1 + \ldots + a_{in} x_n \qquad (i = 1, \ldots, h),$$

where the a's depend only upon the constants of multiplication γ_{ijk}. Hence the roots of the characteristic equation of $x + x'$ are

$$a_{i1}(x_1 + x_1') + \ldots = \omega_i + \omega_i'.$$

* With coefficients in F, in case we are treating algebras over a field F. Since the ω's are now in F and since the work in §§ 36—39 is purely rational, we obtain normalized units (§ 39) with coordinates in F; cf. ex. in §§ 26, 33.

† G. Scheffers, *Math. Ann.*, vol. 39 (1891), p. 305: *Quaternionsysteme, Nicht-quaternionsysteme*. Th. Molien, *ibid.*, vol. 41 (1893), p. 83.

If x and x' are nilpotent, each $\omega_i = \omega_i' = 0$, and $x + x'$ is nilpotent.

By § 32, any number of A_1 is of the form

$$x_1\epsilon_1 + \ldots + x_h\epsilon_h + y_1\eta_1 + \ldots + y_k\eta_k \qquad (66),$$

where the ϵ's are the partial moduli and the η's are nilpotent numbers, each with a definite character. Since

$$(x_1\epsilon_1 + \ldots + x_h\epsilon_h)\,\epsilon_i = x_i\epsilon_i,$$

the characteristic equation of (66) has the roots $x_1, \ldots, x_h$ and hence (66) *is nilpotent if and only if* $x_1, \ldots, x_h$ *are all zero.*

37. Theorem. *The product of any nilpotent number of* A_1 *by any number* u *of* A_1 *is nilpotent.*

It suffices to prove that $\eta_1 u, \ldots, \eta_k u$ are nilpotent.

Consider an η of character (1, 1), and express u as a sum of components each of a definite character. The product of η by a component will be zero or of character $(1, j)$, $j > 1$, and hence a nilpotent number (§ 32), unless the component be of character (1, 1). In the latter case also, the product is nilpotent (§ 34).

Consider an η of character $\neq (i, i)$, say (1, 2). As before it suffices to take u of character (2, 1). Then ηu is of character (1, 1) and

$$\eta u = a\epsilon_1 + \zeta,$$

where a is a scalar and ζ a nilpotent number of character (1, 1). If $a = 0$, the theorem is proved. Henceforth, let $a \neq 0$. Then (§ 34)

$$(\eta u)^2 = a^2\epsilon_1 + \zeta_1, \quad \zeta_1 = 2a\zeta + \zeta^2,$$

where ζ_1 is a nilpotent number of Σ_1 in § 28. By induction,

$$(\eta u)^m = a^m\epsilon_1 + \zeta_m,$$

where ζ_m is nilpotent. If $u\eta$ is nilpotent, its mth power is zero, where m is a certain integer. Then

$$0 = \eta\,(u\eta)^m = (\eta u)^m\eta = a^m\eta + \zeta_m\eta,$$

since $\epsilon_1\eta = \eta$, η being of character (1, 2). Thus $-a^m$ is a root of the characteristic equation of ζ_m. This is impossible since $a \neq 0$ and ζ_m is nilpotent. Hence $u\eta$ is not nilpotent:

$$u\eta = a'\epsilon_2 + \zeta',$$

where $a' \neq 0$ and ζ' is nilpotent of character (2, 2).

If $v \neq 0$ is of character (2, 1), then $\eta v \neq 0$. For, if $\eta v = 0$,

$$0 = u\,(\eta v) = (u\eta)\,v = a'v + \zeta' v,$$

whereas $-a'$ is not a root of the characteristic equation of the nilpotent number ζ' since a' is not zero. If $v_1, \dots, v_t$ form a complete* set of linearly independent numbers of character (2, 1), then $\eta v_1, \dots, \eta v_t$ are linearly independent and of character (1, 1). Indeed, $\Sigma c_i \eta v_i = 0$ implies $\eta v = 0$, where $v = \Sigma c_i v_i$.

If $w \neq 0$ is of character (1, 1), then $uw \neq 0$. For, if $uw = 0$,

$$0 = (\eta u) w = aw + \zeta w,$$

whereas $-a$ is not a root of the characteristic equation of the nilpotent number ζ. If $w_1, \dots, w_\tau$ form a complete set of linearly independent numbers of character (1, 1), then $uw_1, \dots, uw_\tau$ are linearly independent and of character (2, 1). Indeed, $\Sigma c_i uw_i = 0$ implies $uw = 0$, $w = \Sigma c_i w_i$.

By the first result, $t \leqq \tau$; by the second, $\tau \leqq t$. Hence $\tau = t$ and $\eta v_1, \dots, \eta v_t$ form a complete set of linearly independent numbers of character (1, 1). Hence there exist scalars d_i such that $\Sigma d_i \eta v_i = \epsilon_1$. Thus $\eta v = \epsilon_1$ for $v = \Sigma d_i v_i$. Here η is of character (1, 2) and v of character (2, 1). Hence

$$(\eta + v)(\epsilon_1 + v) = \epsilon_1 + v.$$

The characteristic equation of $\eta + v$ has therefore the root unity, whereas it is the sum of two nilpotent numbers of A_1 and hence is nilpotent. The assumption that $a \neq 0$ has therefore led to a contradiction.

By reversing the order of the factors in our products, we see that *the product of any number of A_1 by any nilpotent number is nilpotent.*

COROLLARY. *The nilpotent numbers of A_1 form a linear associative algebra N without a modulus.*

If it had a modulus η, then would $\eta u = u$, whereas unity is not a root of the characteristic equation of the nilpotent number η.

An algebra is called *nilpotent* if all of its numbers are nilpotent.

38. Normalized units of a nilpotent algebra.

THEOREM. *In a nilpotent algebra N, linearly independent units $\eta_1, \dots, \eta_k$, each of a definite character†, can be chosen so that the product of any two units η_i and η_j is a linear function of the units η whose subscripts exceed both i and j.*

If $k = 1$, $\eta_1^2 = a\eta_1$, where a is scalar. Since a is a root of the

* Any number of character (2, 1) is to be a linear function of $v_1, \dots, v_t$ with scalar coefficients. The basis of the definition is the result at the end of § 25.

† This supplement concerning character applies to algebras N contained in an algebra with partial moduli.

characteristic equation of the nilpotent number η_1, we have $a=0$. The theorem is thus true for $k=1$, since it states that η_1^2 is a linear function of the (non-existing) η's whose subscripts exceed 1, and hence that $\eta_1^2=0$.

Assuming that the theorem holds for nilpotent algebras with fewer than k units, we shall prove it for an algebra N with k units $e_1, \ldots, e_k$.

LEMMA. *There exists a number u of N for which*

$$e_1 u=0, \quad e_2 u=0, \ldots, \quad e_k u=0 \tag{67}$$

Among all of the numbers u (including certainly a power of e_1) for which $e_1 u=0$, there may be some making $e_2 u=0$, some making also $e_3 u=0$, etc. Suppose that the Lemma is false. Then none of these u's satisfy (67). Hence there is an integer $m<k$ for which

$$e_1 u=0, \quad e_2 u=0, \ldots, \quad e_m u=0 \tag{68}$$

have a common solution $u \neq 0$ and such that no common solution u makes $eu=0$, where e is any number of N, linearly independent of $e_1, \ldots, e_m$. Any linear combination of solutions of (68) is a solution. Hence we may assume that $u_1, \ldots, u_n$ form a complete set of linearly independent solutions of (68). For $j \leqq m$, $(e_i e_j)\,u$ is zero for every solution u. Hence $e_i e_j$ is a linear function of $e_1, \ldots, e_m$. Thus $e_1, \ldots, e_m$ are units of a nilpotent sub-algebra N_1. Since $m<k$, the hypothesis for our induction shows that our theorem holds for N_1. Hence, if $i \leqq m$, $j \leqq m$, $e_i e_j$ is a linear function of the e's whose subscripts exceed i and j, but not m.

Let ν be any number $\neq 0$ of N. *Either ν is a linear combination of $u_1, \ldots, u_n$, or there exists a number η of N_1 such that $\eta\nu$ is a linear combination $\neq 0$ of $u_1, \ldots, u_n$.* For, if $i \leqq m$, $e_i e_m=0$, as just proved, so that $e_i(e_m \nu)=0$ and $e_m \nu$ is a linear combination of the independent solutions $u_1, \ldots, u_n$ of (68). If $e_m \nu \neq 0$, our italicized statement is proved. If $e_m \nu=0$, every $e_i e_{m-1} \nu=0$ $(i \leqq m)$, since $e_i e_{m-1}$ is a multiple of e_m, as just proved. Thus $e_{m-1}\nu$ is a linear combination of $u_1, \ldots, u_n$. Hence our italicized statement follows unless possibly each $e_i \nu=0$; but then ν itself is a linear combination of $u_1, \ldots, u_n$.

Taking $e_{m+1} u_1$ as ν, our italicized result shows that either $e_{m+1} u_1$ or some η times it is a linear function $\neq 0$ of $u_1, \ldots, u_n$. In the respective cases take $\zeta_1=e_{m+1}$ or ηe_{m+1}. Thus $\zeta_1 u_1 \neq 0$. Since ζ_1 is nilpotent, $\zeta_1 u_1 \neq s u_1$, where s is scalar. Hence, by a change of notation of the u's, we may take $\zeta_1 u_1=u_2$.

For this new u_2, take $e_{m+1} u_2$ as ν in our italicized result. Thus

$\zeta_2 = e_{m+1}$ or $\eta' e_{m+1}$ has the property that $\zeta_2 u_2$ is a linear combination $\neq 0$ of $u_1, \ldots, u_n$. If $\zeta_2 u_2$ be a linear function of u_1, u_2, we have

$$\zeta_1 u_1 = u_2, \ \zeta_2 u_2 = u_3, \ldots, \ \zeta_{p-1} u_{p-1} = u_p, \ \zeta_p u_p = \lambda_1 u_1 + \ldots + \lambda_p u_p \quad (69),$$

for $p = 2$, where the scalars $\lambda_1, \ldots, \lambda_p$ are not all zero. In the contrary case, we may take $\zeta_2 u_2 = u_3$. Proceeding with u_3, we obtain a number ζ_3 such that $\zeta_3 u_3$ is a linear function $\neq 0$ of $u_1, \ldots, u_n$. If it involves only u_1, u_2, u_3, we have (69) for $p = 3$. If not, we take $\zeta_3 u_3 = u_4$, etc.

We may set $\lambda_1 = 0, \ldots, \lambda_{s-1} = 0, \lambda_s \neq 0$, where s is a certain integer $\geqq 1$. Set

$$l = \zeta_p \zeta_{p-1} \ldots \zeta_{s+1} \zeta_s - \lambda_p \zeta_{p-1} \ldots \zeta_s - \ldots - \lambda_{s+2} \zeta_{s+1} \zeta_s - \lambda_{s+1} \zeta_s.$$

Multiply by u_s on the right. By (69) the product is

$$\zeta_p u_p - \lambda_p u_p - \ldots - \lambda_{s+2} u_{s+2} - \lambda_{s+1} u_{s+1} = \lambda_s u_s.$$

Thus $l u_s = \lambda_s u_s$. But l is nilpotent and $\lambda_s \neq 0$. Hence the supposition that the lemma is false has led to a contradiction.

Let therefore $u_1, \ldots, u_n \ (n \geqq 1)$ form a complete set of linearly independent solutions u of the system of equations (67). Let η be any number of N. Then $(e_j u_i)\eta = 0$, so that $u_i \eta$ is a linear function of $u_1, \ldots, u_n$:

$$u_i \eta = a_{i1} u_1 + \ldots + a_{in} u_n.$$

If $u = \Sigma c_i u_i$, the conditions for $u\eta = \rho u$ are

$$\sum_{i=1}^{n} c_i a_{i1} - c_1 \rho = 0, \ \ldots, \quad \sum_{i=1}^{n} c_i a_{in} - c_n \rho = 0.$$

Let ρ be a root of

$$\begin{vmatrix} a_{11} - \rho & a_{21} \ldots & a_{n1} \\ \ldots & \ldots & \ldots \\ a_{1n} & a_{2n} \ldots & a_{nn} - \rho \end{vmatrix} = 0.$$

Then the determinant of the coefficients of $c_1, \ldots, c_n$ in our n equations is zero, which therefore have a set of solutions $c_1, \ldots, c_n$ not all zero. By $u\eta = \rho u$, ρ is a root of the second characteristic equation of the nilpotent number η. Hence $\rho = 0$. Given any η in N, we can therefore find a linear combination $u \neq 0$ of $u_1, \ldots, u_n$ such that $u\eta = 0$.

Apply this result to $\eta = e_1$. Hence there is a linear function $u \neq 0$ of $u_1, \ldots, u_n$ for which $ue_1 = 0$. Some of the solutions u of the latter may make $ue_2 = 0$, some also $ue_3 = 0$, etc. Hence there exists an integer $m \leqq k$ such that

$$ue_1 = 0, \quad ue_2 = 0, \ldots, \quad ue_m = 0 \quad (70)$$

have a common solution u which is a linear function $\neq 0$ of $u_1, \ldots, u_n$, but no such solution making also $ue = 0$, where e is a number of N, linearly independent of $e_1, \dots, e_m$. By a change of notation of $u_1, \dots, u_n$, we may assume that $u_1, \dots, u_{n'}$ form a complete set of linearly independent solutions of (70).

We shall assume that $m < k$ and prove that a contradiction results. For $i \leqq m$, $u(e_i e_j) = 0$ for every solution u of (70). Hence $e_i e_j$ is a linear function of $e_1, \dots, e_m$, which are therefore the units of a sub-algebra N_1, to which our theorem applies. If ν is any number $\neq 0$ of N, *either* ν *is a linear combination of* $u_1, \dots, u_{n'}$ *or there exists a number* η *of* N_1 *such that* $\nu\eta$ *is a linear function* $\neq 0$ *of* $u_1, \dots, u_{n'}$. For, if $i \leqq m$, $e_m e_i = 0$, $(\nu e_m) e_i = 0$ and νe_m is a linear combination of $u_1, \dots, u_{n'}$. If $\nu e_m \neq 0$, the italicized statement holds. If zero, $\nu e_{m-1} e_i = 0$ $(i \leqq m)$, since $e_{m-1} e_i$ is a multiple of e_m. Thus νe_{m-1} is a linear function of $u_1, \dots, u_{n'}$. Taking $u_1 e_{m+1}$ as ν, we obtain, as in the proof of the lemma, a ζ_1 for which $u_1\zeta_1 \neq s u_1$, thence $u_1 \zeta_1 = u_2$, etc. We obtain (69) with $u_i \zeta_i$ in place of $\zeta_i u_i$, and then the contradiction $u_s l = \lambda_s u_s$. Hence $m = k$, so that there exists a solution $u \neq 0$ of (67) and (70), whence $ut = tu = 0$ for every number t in the algebra N.

This u is a sum of numbers u' of definite characters. If η is of character (α, β), $u\eta$ is a sum of products $u'\eta$ each zero or of a definite character. Since $u\eta = 0$, each $u'\eta = 0$. Similarly, each $\eta u' = 0$.

We introduce as new linearly independent units $\eta_1, \eta_2, \dots, \eta_k$, where η_k is one of the u', such that each η has a definite character. Then

$$\eta_i \eta_k = \eta_k \eta_i = 0 \qquad (i = 1, \dots, k) \qquad (71),$$

$$\eta_i \eta_j = \sum_{s=1}^{k} \gamma_{ijs} \eta_s \qquad (i, j = 1, \dots, k).$$

Consider an algebra with the units $\zeta_1, \dots, \zeta_{k-1}$ such that

$$\zeta_i \zeta_j = \sum_{s=1}^{k-1} \gamma_{ijs} \zeta_s \qquad (i, j = 1, \dots, k-1).$$

For λ, μ, ν all less than k, $(\zeta_\lambda \zeta_\mu)\zeta_\nu = \zeta_\lambda(\zeta_\mu \zeta_\nu)$. Indeed, $\eta_\lambda \eta_\mu = \eta' + c\eta_k$, where η' is the same linear function of $\eta_1, \dots, \eta_{k-1}$ that $\zeta_\lambda \zeta_\mu$ is of $\zeta_1, \dots, \zeta_{k-1}$. But $(\eta_\lambda \eta_\mu)\eta_\nu = \eta'\eta_\nu$, so that its part free of η_k is the same function of $\eta_1, \dots, \eta_{k-1}$ that $(\zeta_\lambda \zeta_\mu)\zeta_\nu$ is of $\zeta_1, \dots, \zeta_{k-1}$. Similarly, the part of $\eta_\lambda(\eta_\mu \eta_\nu)$ free of η_k is the same function of $\eta_1, \dots, \eta_{k-1}$ that $\zeta_\lambda(\zeta_\mu \zeta_\nu)$ is of $\zeta_1, \dots, \zeta_{k-1}$. Since the products of the η's are equal, those of the ζ's are. Hence the ζ's are the units of a linear associative algebra.

Since $\gamma_{iks} = 0$ $(i, s = 1, \dots, k)$ by (71_1), the elements of the kth row of the characteristic determinant (40) of the η-algebra are $0, \ldots, 0, -\omega$. Since $\gamma_{kjs}=0\,(j, s=1, \ldots, k)$ by (71_2), the minor of $-\omega$ is the characteristic determinant of the ζ-algebra. Hence every number of the latter is nilpotent. Since the theorem applies to this ζ-algebra, we may suppose that new ζ's have been introduced such that, if i, j, s are less than k, $\gamma_{ijs}=0$ unless $s>i$, $s>j$. As just noted, this result holds also if $i=k$ or $j=k$. Hence $\eta_1, \ldots, \eta_k$ satisfy the requirements of our theorem.

39. Normalized units of an algebra A_1 of the first category. In view of the last theorem, the result at the end of § 25, and the remark in § 36 on the units, we have the

THEOREM*. *If in the characteristic equation $\delta(\omega)=0$ for A_1, ω is an h-valued function of $x_1, \ldots, x_n$, we may choose as normalized units $\epsilon_1, \ldots, \epsilon_h$, $\eta_1, \ldots, \eta_k$, where each η has a definite character, $\epsilon_i^2=\epsilon_i$, $\epsilon_i\epsilon_j=0\,(i\neq j)$, while $\eta_a\eta_b$ is a linear function of those η's whose subscripts exceed a and b and have the same character as $\eta_a\eta_b$.*

Conversely, any such algebra is of the first category.

To prove the converse, arrange the n units in the following order: first ϵ_1; then the units η_i of character $(1, 1), (2, 1), \ldots, (h, 1)$ in ascending order of their subscripts i; then ϵ_2 and the units η_i of character $(1, 2), (2, 2), \ldots, (h, 2)$ in the order of their subscripts i; etc. Let $\zeta_1, \ldots, \zeta_n$ be the units thus arranged. Let $n_{\alpha\beta}$ be the number of units η of character (α, β) and set

$$n_\beta = 1 + n_{1\beta} + n_{2\beta} + \ldots + n_{h\beta}.$$

The characteristic determinant of $z=z_1\zeta_1+\ldots+z_n\zeta_n$ is (40):

$$\delta(\omega) \equiv \begin{vmatrix} \Sigma z_i\gamma_{i11}-\omega & \Sigma z_i\gamma_{i21}\ldots\Sigma z_i\gamma_{in1} \\ \ldots\ldots\ldots\ldots & \ldots\ldots\ldots\ldots \\ \Sigma z_i\gamma_{i1n} & \Sigma z_i\gamma_{i2n}\ldots\Sigma z_i\gamma_{inn}-\omega \end{vmatrix} \qquad (72).$$

The elements which lie simultaneously in one of the first n_1 rows and last $n-n_1$ columns are zero, since

$$\gamma_{i\lambda\mu}=0 \quad (i=1, \ldots, n;\ \lambda>n_1;\ \mu\leqq n_1).$$

Indeed, for such a set of subscripts, ζ_μ is of character $(\bullet 1)$, ζ_λ of character $(\bullet\delta)$, $\delta>1$, $\zeta_i\zeta_\lambda$ is zero or of character $(\bullet\delta)$. Hence in

$$\zeta_i\zeta_\lambda = \sum_{\mu=1}^{n} \gamma_{i\lambda\mu}\zeta_\mu \qquad (73),$$

* G. Scheffers, *Math. Ann.*, vol. 39 (1891), p. 293. He took a definition of algebras A_1 different from the definition we have used following Cartan.

ζ_μ $(\mu \leqq n_1 < \lambda)$ does not occur. Hence its coefficient is zero. Thus $\delta(\omega)$ has as a factor the minor whose elements lie in the first n_1 rows and first n_1 columns. Similarly, it has n_2-rowed, ..., n_h-rowed minors as factors.

In the n_1-rowed minor all of the elements above the main diagonal are zero, while the element in the λth row and main diagonal is $x_t - \omega$, where $(t, 1)$ is the character of ζ_λ. First, let $1 < \lambda \leqq n_1$, $1 < \mu \leqq n_1$, so that ζ_λ, ζ_μ are η's. Then if ζ_i is an η, $\gamma_{i\lambda\mu} = 0$ if $\mu \leqq \lambda$, by the above theorem*. But if $\zeta_i = \epsilon_t$ and ζ_λ is of character $(a, 1)$, $\zeta_i\zeta_\lambda = 0$ or ζ_λ, according as $a \neq t$, $a = t$, so that $\gamma_{i\lambda\mu} = 0$ if $\mu \neq \lambda$, and $\gamma_{i\lambda\lambda} = 0$ or 1, according as $a \neq t$ or $a = t$. Second, let $\mu = 1$, so that $\zeta_\mu = \epsilon_1$. There is no ϵ_1 in the sum (73) if ζ_i and ζ_λ are both η's. Next, let $\zeta_i = \epsilon_t$, so that $\zeta_i\zeta_\lambda = 0$ or ζ_λ, according as ζ_λ is not or is of character $(t\bullet)$; thus ϵ_1 occurs in the sum (73) only when $\lambda = t = 1$. Since $\epsilon_1\epsilon_1 = \epsilon_1$, $\gamma_{i11} = 0$ $(i > 1)$, $\gamma_{111} = 1$. Finally, if ζ_λ $(\lambda \leqq n_1)$ is an ϵ, it is ϵ_1, and the sum (73) contains ϵ_1 only when it reduces to the preceding case $\epsilon_1\epsilon_1 = \epsilon_1$. Third, if $\lambda = 1$, $\gamma_{i\lambda\mu}$ is below the main diagonal unless $\mu = 1$. Hence in our minor, all elements above the main diagonal are zero and all in the diagonal are zero except the $\gamma_{i\lambda\lambda}$ for which $\zeta_i = \epsilon_t$ and ζ_λ is of character $(t, 1)$, and these γ's are unity. Hence if z be given the former notation (66) we have the result at the beginning of this paragraph. Thus the n_1-rowed minor equals

$$(x_1 - \omega) \prod_{t=1}^{h} (x_t - \omega)^{n_{t1}},$$

the first factor coming from ϵ_1. Similarly,

$$(x_2 - \omega) \prod_{t=1}^{h} (x_t - \omega)^{n_{t2}}$$

is the value of the n_2-rowed minor, etc. Set

$$n_t' = 1 + n_{t1} + n_{t2} + \ldots + n_{th}.$$

Hence†

$$\delta(\omega) = (x_1 - \omega)^{n_1'} (x_2 - \omega)^{n_2'} \ldots (x_h - \omega)^{n_h'} \qquad (74).$$

For example, if the units are ϵ_1, η_1, ϵ_2, η_2, where η_1 is of character (1, 1), η_2 of character (1, 2), we have an associative algebra for which the characteristic determinant of $x_1\epsilon_1 + y_1\eta_1 + x_2\epsilon_2 + y_2\eta_2$ is

* The ζ_λ and hence all of the units of the sum (73) are η's of character $(\bullet 1)$, the relative order of which was not changed by our rearrangement of the units.

† The conclusions of Cartan (*l.c.*, §§ 39, 40) are erroneous.

$$\delta(\omega)=\begin{vmatrix} x_1-\omega & 0 & 0 & 0 \\ y_1 & x_1-\omega & 0 & 0 \\ 0 & 0 & x_2-\omega & 0 \\ 0 & 0 & y_2 & x_1-\omega \end{vmatrix}=(x_1-\omega)^3(x_2-\omega).$$

As a check, note that $n_1'=3$, $n_2'=1$, by inspection of the units.

Since the roots are linear functions of the x's, the algebra is of the first category. Since $x_1, \ldots, x_h$ are independent variables and since $\delta(\omega)$ is a covariant (§§ 16, 17), we conclude that $n_1', \ldots, n_h'$ are unaltered by a linear transformation of units which replaces algebra A_1 by one with normalized units*. By using the order (1, 1), (1, 2), ..., (1, h); (2, 1), ..., (2, h), ..., we see that the exponents $n_1, \ldots, n_h$ in $\delta'(\omega)$ are unaltered by transformation.

Moreover, each $n_{\alpha\beta}$ has the last property (see Cartan, *l.c.*, p. 36).

There is an extensive list of papers relating to the determination and classification of algebras of the first category (*Encyc. Sc. Math.*, vol. I, 1, p. 425).

ALGEBRAS A_2 OF THE SECOND CATEGORY.

40. Properties of the characteristic determinant for A_2.

We shall employ the normalized units ϵ_i, η_g of § 32. A linear combination of the nilpotent units $\eta_1, \ldots, \eta_k$ need not be nilpotent.

Arrange the units in the order $\zeta_1, \ldots, \zeta_n$ as in § 39. The first property of determinant (72) holds also here: $\delta=\delta_1\delta_2\ldots\delta_h$, where δ_1 is the minor composed of the elements in the first n_1 rows and columns of δ, etc.

In (73) set $i=1$, $\lambda \leqq n_1$. Then ζ_λ is of character (•1). Since ζ_1 is the partial modulus ϵ_1, we have $\zeta_1\zeta_\lambda=0$ or ζ_λ according as the character of ζ_λ is $(\alpha, 1)$, $\alpha>1$ or (1, 1). Hence $\gamma_{1\lambda\mu}=0$ unless $\mu=\lambda$ and ζ_λ is of character (1, 1). Thus z_1 occurs only in the main diagonal of δ_1 and there only in the first $n_{11}+1$ terms (with coefficient unity). Similarly, z_{n_1+1}, the coefficient of ϵ_2, occurs only in the main diagonal and

* But under the transformation each ϵ_i may be increased by a properly chosen nilpotent number. For example, if $h=2$, $k=1$, and $\eta_1=\eta$ is of character (1, 2), the multiplication table is

$$\epsilon_1^2=\epsilon_1, \quad \epsilon_2^2=\epsilon_2, \quad \epsilon_1\epsilon_2=\epsilon_2\epsilon_1=\epsilon_2\eta=\eta\epsilon_1=0, \quad \eta\epsilon_2=\eta, \quad \epsilon_1\eta=\eta, \quad \eta^2=0.$$

This is unaltered if we replace ϵ_1 by $\epsilon_1'=\epsilon_1+\eta$, ϵ_2 by $\epsilon_2'=\epsilon_2-\eta$.

Various writers have overlooked this possibility (see end of § 61).

there only in n_{21} terms (with coefficient unity), etc. Using the second notation (66) for z, we see that $x_1, x_2, \ldots, x_h$ occur only in the main diagonal of δ_1 and that each element of the diagonal contains one and but one of these x's (with coefficient unity). Thus every element of δ_1 is linear and homogeneous in

$$x_1' = x_1 - \omega, \ldots, x_h' = x_h - \omega, y_1, \ldots, y_k \qquad (75).$$

Hence δ_1 is homogeneous in $x_1', \ldots, y_k$.

For brevity we shall say that y_i is of the same character as η_i and x_i' of character (i, i). If, in (73), ζ_i is of character $(\alpha\bullet)$ then $\gamma = 0$ unless ζ_μ is of character $(\alpha\bullet)$. Conversely, if ζ_μ is of character $(\alpha\bullet)$ then $\gamma = 0$ unless ζ_i is of character $(\alpha\bullet)$. Hence the variables z_i of character $(\alpha\bullet)$ enter the same rows of the determinant δ_1. Hence δ_1 is homogeneous and of degree $n_{\alpha 1}$ or $n_{\alpha 1} + 1$ in the z_i of character $(\alpha\bullet)$, according as $\alpha > 1$ or $\alpha = 1$.

Similarly δ_1 is homogeneous and of degree $n_{\alpha 1}$ or $n_{\alpha 1} + 1$ in the variables of character $(\bullet\alpha)$, according as $\alpha > 1$ or $\alpha = 1$.

Let $\delta(\omega) = P_1^{a_1} \ldots P_l^{a_l}$, where $P_1, \ldots, P_l$ are distinct irreducible functions of their arguments (75), the degree of P_i in ω being d_i. Then $d_1 + \ldots + d_l = h$, if $\delta(\omega) = 0$ defines ω as an h-valued function of the x's and y's. For $y_1 = 0, \ldots, y_k = 0$, we saw that $\delta(\omega)$ reduces to the product (74) of its diagonal elements. Thus P_i reduces to a product of d_i factors $x_j - \omega$. The resulting Σd_i linear factors are $x_1 - \omega, \ldots, x_h - \omega$ in some order. Hence no one appears twice in the same P_i or appears in different P's. We may thus set

$$(P_1)_{y=0} = (x_1 - \omega) \ldots (x_p - \omega) \equiv x_1' \ldots x_p' \qquad (76).$$

The factor δ_1 of δ is homogeneous in the variables of character $(\alpha\bullet)$. Hence each irreducible factor P_1 of δ_1 is homogeneous in these variables. To find its degree in them, it suffices to examine one term (76), which is linear and homogeneous in the variables of character $(\alpha\bullet)$. We have therefore the

THEOREM. *The irreducible factor P_1 of the determinant δ_1 is linear and homogeneous in those of its arguments* (75) *whose characters are* $(1\bullet)$, *linear and homogeneous in those of characters* $(2\bullet), \ldots,$ *and in those of characters* $(p\bullet)$; *likewise for the variables of characters* $(\bullet 1), \ldots,$ *and for* $(\bullet p)$. *No further variables occur in* P_1.

41. Example. In the matric algebra of four units e_{ij} of § 4, e_{ij} is of character (i, j). The characteristic determinant of

$$x_{11}e_{11} + x_{12}e_{12} + x_{21}e_{21} + x_{22}e_{22}$$

is, by § 35, the square of

$$\delta_1 = \begin{vmatrix} x_{11}' & x_{12} \\ x_{21} & x_{22}' \end{vmatrix}, \quad x_{11}' = x_{11} - \omega, \quad x_{22}' = x_{22} - \omega.$$

Hence $P_1 = \delta_1$ is linear and homogeneous in x_{11}' and x_{12} of characters ($1\bullet$), also in x_{21} and x_{22}' of characters ($2\bullet$); also in x_{11}' and x_{21} of characters ($\bullet 1$); and in x_{12}, x_{22}' of characters ($\bullet 2$).

42. Sub-algebras S_i of A_2. To obtain P_1 we need consider only numbers involving the units of characters (α, β), $\alpha, \beta = 1, \ldots, p$. These are the units of a sub-algebra S_1.

By the argument leading to (76), we may set

$$(P_2)_{y=0} = x'_{p+1}\, x'_{p+2} \ldots x'_q.$$

To obtain P_2 we need consider only numbers involving the units of characters (α, β), $\alpha, \beta = p+1, \ldots, q$; these are the units of a sub-algebra S_2, etc. The product of a number of S_1 by one of S_2 is zero, and *vice versa.* In the characteristic determinant of the reducible (§ 21) algebra $S_1 + S_2$, the elements in the first and third "quadrants" are all zero, so that it equals the product of the characteristic determinants of S_1 and S_2. Similarly, for $S_1 + S_2 + \ldots + S_l$. Thus $\delta(\omega)$ is unaltered when we replace by zero each variable not corresponding to a unit of S_1, $S_2, \ldots,$ or S_l. Hence the characteristic determinant of S_i is $P_i^{a_i}$.

For the moment, let $[\alpha, \beta]$ denote a variable (75) of character (α, β). Let P_1 be the irreducible factor of degree $p > 1$ of $\delta(\omega)$, with the properties given in the theorem of § 40. It involves a variable of character (α, β), $\alpha \neq \beta$. Otherwise, a power of P_1 would be the characteristic determinant of a number of algebra S_1 having each $[\alpha, \beta] = 0$, $\alpha \neq \beta$, and hence of the direct sum of the algebras $\Sigma_1, \ldots, \Sigma_p$ of numbers of characters $(1, 1), \ldots, (p, p)$, respectively. But (§ 42), P_1 would then be a product of a function of the variables $[1, 1]$, by a function of the $[2, 2]$, etc., and hence a product of linear functions, whereas it is irreducible and of degree > 1.

By choice of the notation, we may assume that P_1 has a term with the factor $[1, 2]$ and hence a factor $[2, j]$, $j \neq 2$, which we may take to be $[2, 1]$ or $[2, 3]$; in the latter case, a factor $[3, 1]$ or $[3, 4]$, etc. Thus P_1 has a term

$$[1, 2][2, 3] \ldots [\alpha - 1, \alpha][\alpha, 1] \bullet [\alpha + 1, \alpha + 2] \ldots [\beta, \alpha + 1] \ldots \prod_{i=\gamma+1}^{p} [i, i] \qquad (77),$$

in which $[i, i]$ for $i = \lambda + 1, \ldots, p$ are variables $x_i' = x_i - \omega$, while the remaining $[i, i]$ are variables y_i. Set $x_1 = 0, \ldots, x_p = 0$. Then (77) becomes a multiple of $\omega^{p-\lambda}$, while the term (76) of P_1 reduces to $\pm\omega^p$. Hence algebra S_1 contains a number whose characteristic equation has a root $\neq 0$, the number being a sum of numbers of characters

$$(1, 2), (2, 3), \ldots, \quad (a, 1), (a+1, a+2), \ldots, \quad (\beta, a+1), \quad (i, i) \qquad (i=\gamma+1, \ldots, \lambda) \qquad (78),$$

those of character (i, i) being nilpotent.

43. Preliminary criterion for an algebra A_2 of the second category. A necessary and sufficient condition that an algebra be of the second category is that it contain a number

$$u = (e_{12} + e_{23} + \ldots + e_{a-1a} + e_{a1}) + (e_{a+1a+2} + \ldots + e_{\beta a+1}) + \ldots + \sum_{i=\gamma+1}^{\lambda} \eta_{ii},$$

whose characteristic equation has a root not zero, where the e_{ij} and η_{ii} are linear functions of nilpotent units $\eta_1, \ldots, \eta_k$ of characters (i, j) and (i, i), respectively, and where one of the integers $a, \beta - a, \ldots, \lambda - \gamma$ is positive.

The condition is necessary by § 42, and sufficient by § 36 [see § 47].

44. Since u is not nilpotent, there is a number $v \neq 0$ such that $uv = \omega v$, ω a scalar $\neq 0$. Take u/ω as a new u. Then $uv = v$. Hence if v be expressed as a sum of numbers each of a definite character (r, s), then $r \leqq \lambda$. Let $v_{\gamma+1}$ be the sum of the parts of v of character $(\gamma+1 \bullet)$. Then

$$uv_{\gamma+1} = \eta_{\gamma+1\gamma+1} v_{\gamma+1} = v_{\gamma+1} = 0,$$

since the η is nilpotent. Hence each term $\neq 0$ of v is of character (r, s), $r \leqq \gamma$. After a change of notation of the cycles of u, we may assume that v has a term $\neq 0$ of character $(1 \bullet)$. Call v' the number $\neq 0$ which is the sum of the terms of v of characters $(1 \bullet) \ldots, (a \bullet)$. Set $u' = e_{12} + e_{23} + \ldots + e_{a1}$. Then $u'v' = v'$. Drop the accents. We may therefore assume that initially

$$u = e_{12} + e_{23} + \ldots + e_{a1}, \quad uv = v,$$

$$v = \sum_{i=1}^{h} (v_{1i} + v_{2i} + \ldots + v_{ai}) \neq 0,$$

where v_{ai} is zero or of character (a, i). By $uv = v$,

$$e_{12}v_{2i} = v_{1i}, \quad e_{23}v_{3i} = v_{2i}, \ldots, \quad e_{a1}v_{1i} = v_{ai}.$$

Let i have a fixed value for which one of the $v_{1i}, \ldots, v_{ai}$ is $\neq 0$ and hence all are $\neq 0$. It follows at once that

$$\begin{aligned} e_{12}e_{23}\ldots e_{a-1a}\ e_{a1}\ \ v_{1i} &= v_{1i}, \\ e_{23}e_{34}\ldots e_{a1}\ \ e_{12}\ \ v_{2i} &= v_{2i}, \\ &\ldots\ldots\ldots\ldots \\ e_{a1}e_{12}\ldots e_{a-2a-1}e_{a-1a}v_{ai} &= v_{ai}. \end{aligned}$$

The product of the e's in the first of these equations has unity as a root of its characteristic equation and is of character (1, 1); hence (§ 31) it equals $a\epsilon_1 + \zeta_1$, where a is a scalar and ζ_1 is nilpotent, the characteristic determinant of $a\epsilon_1 + \zeta_1$ being a power of $a - \omega$. Thus $a = 1$,

$$\begin{aligned} e_{12}e_{23}\ldots e_{a-1a}\ e_{a1} &= \epsilon_1 + \zeta_1, \\ &\ldots\ldots\ldots\ldots \\ e_{a1}e_{12}\ldots e_{a-2a-1}e_{a-1a} &= \epsilon_a + \zeta_a, \end{aligned}$$

ζ_i being nilpotent and of character (i, i).

45. Notation. Set $e_{13} = e_{12}e_{23}$, $e_{a2} = e_{a1}e_{12}$, and, in general,

$$e_{ij} = e_{i\,i+1}e_{i+1\,i+2}\ldots e_{j-2\,j-1}e_{j-1\,j} \qquad (i \neq j) \qquad (79),$$

where e_{aa+1} is to be replaced by e_{a1}. The last equations in § 44 become

$$e_{ji}e_{ij} = \epsilon_j + \zeta_j \qquad (i, j = 1, \ldots, a;\ i \neq j) \quad (80).$$

If η is a number $\neq 0$ of character (j, λ), then $e_{ij}\eta \neq 0$. For, if zero,

$$0 = e_{ji}e_{ij}\eta = \eta + \zeta_j\eta,$$

whereas -1 is not a root of the characteristic equation of the nilpotent number ζ_j. Let $n_1, \ldots, n_t$ be a complete set of linearly independent numbers of character (j, λ). Then $e_{ij}n_1, \ldots, e_{ij}n_t$ are linearly independent numbers of character (i, λ). For, if $\Sigma c_s e_{ij} n_s = 0$, then $e_{ij}\eta = 0$ when $\eta = \Sigma c_s n_s$. If the resulting $e_{ij}n_s$ did not form a complete set, the n_s would not. *Hence a complete set of linearly independent numbers of character (i, λ) is obtained from a complete set of character (j, λ) by multiplying e_{ij} by the latter.*

It may be proved in the same manner that *there are as many linearly independent numbers of character (λ, i) as of character (λ, j), and the latter are given by the products of the former by e_{ij}.*

In these theorems, $i, j = 1, \ldots, a;\ \lambda = 1, \ldots, h$.

46. In particular, if $n_1, \ldots, n_t$ form a complete set of linearly independent numbers of character $(a, 1)$, where $a > 1$, then $e_{1a}n_1, \ldots, e_{1a}n_t$ form a complete set of character (1, 1). Hence for suitably chosen scalars c_s, $\Sigma c_s n_s$ is a number $e_{\alpha 1}'$ of character $(\alpha, 1)$ such that $e_{1\alpha} e_{\alpha 1}' = \epsilon_1$, a number of character $(1, 1)$. Expressed otherwise, by means of (79),

$$e_{1i}\beta = \epsilon_1, \quad e_{1i} = e_{12}e_{23} \ldots e_{i-1i}, \quad \beta = e_{ii+1} \ldots e_{\alpha-1\alpha} e_{\alpha 1}'.$$

Since e_{1i} and β are of characters $(1, i)$ and $(i, 1)$, respectively,

$$\beta e_{1i} = a\epsilon_i + \zeta_i',$$

where a is scalar and ζ_i' is nilpotent of character (i, i). Multiplying this on the left by e_{1i} and applying (79), we get

$$e_{1\alpha} e_{\alpha 1}' e_{1i} \equiv e_{1i} = a e_{1i} + e_{1i}\zeta_i'.$$

Thus, by (56), $1 - a$ is a root of the left-hand characteristic equation of the nilpotent number ζ_i'. Hence $a = 1$, $e_{1i}\zeta_i' = 0$. This contradicts the statement after (80), unless $\zeta_i' = 0$. In $\beta e_{1i} = \epsilon_i$, we drop the accent on $e_{\alpha 1}'$ and write e_{ii} for ϵ_i. Thus

$$e_{ii+1} e_{i+1\,i+2} \ldots e_{\alpha-1\alpha} e_{\alpha 1} e_{12} \ldots e_{i-1i} = e_{ii} \quad (i = 1, \ldots, \alpha).$$

Define e_{ij} by (79). Then we get

$$e_{ji} e_{ij} = e_{jj}, \quad e_{ji} e_{il} = e_{jl} \qquad (80').$$

Hence we have α^2 linearly independent numbers $e_{ij}\,(i, j = 1, \ldots, \alpha)$ of character (i, j), which have the same multiplication table (14) as the matric algebra of α^2 units.

THEOREM. *Any algebra of the second category contains a sub-algebra equivalent to the matric algebra of α^2 units, $\alpha > 1$.*

The matric algebra of α^2 units e_{ij} has as a sub-algebra the matric algebra of the β^2 units $e_{ij}\,(i, j = 1, \ldots, \beta)$, if $\beta < \alpha$. By § 13, we have the

COROLLARY. *Any algebra of the second category contains an algebra equivalent to the algebra of complex quaternions.*

47. Normalized units of an algebra A_2 of the second category.

For brevity we shall say that there is a path joining α and β, where α and β are distinct positive integers $\leqq h$, if there is a number $\eta_{\alpha\beta}$ of character (α, β) and a number $\eta_{\beta\alpha}$ of character (β, α) such that $\eta_{\alpha\beta} + \eta_{\beta\alpha}$ is not nilpotent. The existence of a path is a necessary and sufficient condition that an algebra be of the second category. It is sufficient by § 36 and necessary since there is a path joining 1 and 2 in view of

$$(e_{12} + e_{21})(e_{11} + e_{21}) = e_{11} + e_{21}.$$

After rearranging the subscripts 1, ..., h, we may assume that they fall into sets (1, 2, ..., p), ($p+1$, ..., q), ..., such that there is no path joining a number of one set with a number of another set, but that any two given numbers, as 1 and 2, of the same set may be joined by a path; or else other numbers, as 3 and 4, of that set may be found such that there exist paths as 13, 34, 42, securing passage from the first to the second given numbers 1, 2. To secure such passage between any two points, 1, 2, ..., p, evidently $p-1$ paths are necessary and sufficient. They need not form a single broken line, as was assumed by Cartan (*l.c.*, p. 48), who did not employ the suggestive terminology of paths. For $p=5$, the possible figures (trees of Cayley) are as follows:

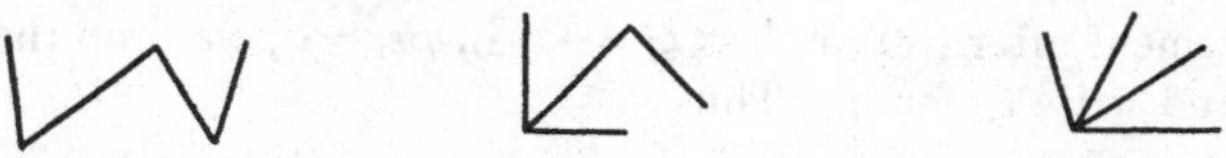

In any case we may assume that there are paths joining 1 with 2 and 2 with 3, and hence numbers $\eta_{12}+\eta_{21}$ and $\eta_{23}+\eta_{32}$ neither nilpotent. Calling the first u, we can find a number $v \neq 0$ such that $uv = v \neq 0$, where v is necessarily a sum of numbers of characters (1•), (2•). Proceeding as in §§ 44—46, with $a=2$, and making a transformation of units (§ 46) which alters at most those of characters (1, 2) and (2, 1), we obtain units e_{12} and e_{21} of those characters such that

$$e_{12}e_{21}=e_{11},\quad e_{21}e_{12}=e_{22}\quad (e_{ii}\text{ partial moduli}) \tag{81}$$

Hence this normalization is not disturbed in treating $\eta_{23}+\eta_{32}$ similarly. Hence we may set also

$$e_{23}e_{32}=e_{22},\quad e_{32}e_{23}=e_{33} \tag{82}$$

If the paths form a single broken line, we get similarly

$$e_{34}e_{43}=e_{33},\ e_{43}e_{34}=e_{44},\ \ldots,\ e_{pp-1}e_{p-1p}=e_{pp} \tag{83}$$

To secure passage from 1 to 3 we must use the paths 1, 2 and 2, 3 and accordingly we set $e_{13}=e_{12}e_{23}$, $e_{31}=e_{32}e_{21}$. In general, we set

$$e_{i\,i+j}=e_{i\,i+1}\,e_{i+1\,i+2}\cdots e_{i+j-1\,i+j},\quad e_{i+j\,i}=e_{i+j\,i+j-1}\cdots e_{i+1\,i}.$$

It is easily seen that the p^2 numbers e_{ij} are the units of a matric algebra (14).

The same result follows if the paths are distributed in any other manner; we have only to define e_{ij} to be the product of the e's whose pairs of subscripts give the uniquely determined series of paths securing passage from i to j. For example, if $p=4$, and if the paths radiate from 2 to 1, 3, 4, respectively, we have (81), (82), and

$$e_{24}e_{42}=e_{22},\quad e_{42}e_{24}=e_{44} \qquad (84),$$

and define the remaining e's by

$$e_{13}=e_{12}e_{23},\quad e_{31}=e_{32}e_{21},\quad e_{14}=e_{12}e_{24},$$

$$e_{41}=e_{42}e_{21},\quad e_{34}=e_{32}e_{24},\quad e_{43}=e_{42}e_{23},$$

so that e_{ij} is of character (i, j) and relations (14) follow.

The theorems of § 45 still hold, with now $i, j=1, \ldots, p$.

Let $\eta_{11}^{(\rho)}$ $(\rho=0, 1, 2, \ldots)$ denote the units of character (1, 1), where $\eta_{11}^{(0)}=e_{11}$. Then all the units of character (i, j), $i \leqq p$, $j \leqq p$, are given by

$$\eta_{ij}^{(\rho)} \equiv e_{i1}\eta_{11}^{(\rho)}e_{1j}.$$

Similarly, if λ is a fixed integer $>p$, and $\eta_{1\lambda}^{(\sigma)}$ $(\sigma=1, 2, \ldots)$ are the units of character $(1, \lambda)$, those of character (i, λ) are

$$\eta_{i\lambda}^{(\sigma)} \equiv e_{i1}\eta_{1\lambda}^{(\sigma)} \quad (\sigma=1, 2, \ldots).$$

We may set

$$\eta_{11}^{(\rho)}\eta_{11}^{(\sigma)}=\sum_\tau \alpha_{\rho\sigma\tau}\eta_{11}^{(\tau)},\quad \eta_{11}^{(\rho)}\eta_{1\lambda}^{(\sigma)}=\sum_\tau \beta_{\rho\sigma\tau}\eta_{1\lambda}^{(\tau)} \qquad (85).$$

Then

$$\eta_{ij}^{(\rho)}\eta_{jl}^{(\sigma)}=\sum_\tau \alpha_{\rho\sigma\tau}\eta_{il}^{(\tau)},\quad \eta_{ij}^{(\rho)}\eta_{j\lambda}^{(\sigma)}=\sum_\tau \beta_{\rho\sigma\tau}\eta_{i\lambda}^{(\tau)}.$$

For example, the second product equals

$$e_{i1}\eta_{11}^{(\rho)}e_{1j}\cdot e_{j1}\eta_{1\lambda}^{(\sigma)}=e_{i1}\eta_{11}^{(\rho)}\eta_{1\lambda}^{(\sigma)}=\sum_\tau \beta_{\rho\sigma\tau}e_{i1}\eta_{1\lambda}^{(\tau)}.$$

Similarly, $\eta_{\lambda i}^{(\rho)}\eta_{ij}^{(\sigma)}$ is a linear function of the $\eta_{\lambda j}^{(\tau)}$ whose coefficients are independent of i, j. Hence, if we are given the multiplication table (85), etc., of the sub-algebra A_1 composed of the units of characters

$$(1, 1), (1, p+1), (1, q+1), \ldots, (p+1, 1), (p+1, p+1), \ldots$$

we can deduce the multiplication table of the main algebra A_2.

In view of the definition of the sets $(1, 2, \ldots, p), \ldots,$

$$\eta_{1\,p+1}+\eta_{p+1\,1}, \ldots$$

are nilpotent, so that the sub-algebra A_1 is of the first category.

We therefore have a simple process to derive all algebras of the second category from those of the first.

Let $E_1, \ldots, E_H, H_1, \ldots, H_K$ be normalized units (§ 39) of A_1:

$$E_i^2=E_i,\quad E_iH_\rho=H_\rho E_j=H_\rho,\quad H_\rho H_\sigma=\Sigma\gamma_{\rho\sigma\tau}H_\tau \qquad (86),$$

where H_ρ is of character (i, j), H_σ of character (j, l), and each H_τ of character (i, l), and $\tau > \rho$, $\tau > \sigma$. *As normalized units* of the algebra A_2 of the second category we take sets of p_i^2 units $e_{\alpha\beta}^{(i)}$ $(\alpha, \beta = 1, \ldots, p_i)$ corresponding to the partial moduli E_i, and sets of $p_i p_j$ units $\eta_{\alpha\beta}^{(\rho)}$ $(\alpha = 1, \ldots, p_i;\ \beta = 1, \ldots, p_j)$ corresponding to H_ρ of character (i, j). If N_{ij} is the number of H_ρ of character (i, j), these*

$$n = \sum_{i=1}^{H} p_i^2 + \sum_{i,j} N_{ij} p_i p_j$$

normalized units of A_2 have the multiplication table

$$e_{\alpha\beta}^{(i)} e_{\beta\gamma}^{(i)} = e_{\alpha\gamma}^{(i)} \qquad (\alpha, \beta, \gamma = 1, \ldots, p_i) \qquad (87),$$

$$e_{\alpha\beta}^{(i)} \eta_{\beta\gamma}^{(\rho)} = \eta_{\alpha\gamma}^{(\rho)},\ \eta_{\beta\gamma}^{(\rho)} e_{\gamma\delta}^{(j)} = \eta_{\beta\delta}^{(\rho)} \qquad (\alpha, \beta \leqq p_i;\ \gamma, \delta \leqq p_j) \qquad (88),$$

$$\eta_{\alpha\beta}^{(\rho)} \eta_{\beta\gamma}^{(\sigma)} = \sum_\tau \gamma_{\rho\sigma\tau} \eta_{\alpha\gamma}^{(\tau)} \qquad (\alpha \leqq p_i,\ \beta \leqq p_j,\ \gamma \leqq p_l,\ \tau > \rho,\ \tau > \sigma) \qquad (89),$$

corresponding to the three types (86). *All products not written are zero*†.

For a simplified statement of part of the content of this theorem, see § 50.

48. Characteristic determinant δ of an algebra A_2. Let the units E, H of the sub-algebra (86) be arranged as in § 39 so that in its characteristic determinant δ_1 every element above the main diagonal is zero. The corresponding sets of units of A_2 are arranged in this order, while the units in the set corresponding to H_ρ of character (i, j) are arranged in the order

$$\eta_{11}^{\rho},\ \eta_{21}^{\rho},\ \ldots,\ \eta_{p_i 1}^{\rho},\ \eta_{12}^{\rho},\ \ldots,\ \eta_{p_i 2}^{\rho},\ \eta_{13}^{\rho},\ \ldots,\ \eta_{p_i p_j}^{\rho},$$

and the subscripts of the $e_{\alpha\beta}^{(i)}$ corresponding to E_i are arranged similarly.

Consider the element in the rth row and cth column of δ_1. Let (i, j) be the character of the rth unit of the sub-algebra, and (l, t) that of the cth unit. Then the matrix of δ is obtained from that of δ_1 by replacing the element in the rth row and cth column by a rectangular array of $p_i p_j$ lines and $p_l p_t$ columns. Since δ_1 is the product of its

* As in § 22, we may express the theorem as follows: Any A_2 can be deduced from an A_1, given by (86), by regarding the coefficient of E_i to be a square matrix of p_i^2 elements and that of H_ρ, of character (i, j), to be a rectangular matrix of p_i rows and p_j columns, these matrices to be regarded as commutative with each E and H.

† Molien, *Math. Ann.*, vol. 41 (1893), p. 83, by use of group theory; Cartan, *l. c.*, as in the text.

diagonal elements, δ equals the product of the determinants of the diagonal arrays, obtained by setting $c = r$, the rth being a $p_i p_j$-rowed determinant. The latter is schematically

$$\begin{vmatrix} M_i & O \ldots & O \\ O & M_i \ldots & O \\ \ldots & \ldots & \ldots \\ O & O & \ldots M_i \end{vmatrix}, \quad M_i \equiv \begin{pmatrix} x_{11}^{(i)} - \omega & x_{12}^{(i)} & \ldots & x_{1p_i}^{(i)} \\ x_{21}^{(i)} & x_{22}^{(i)} - \omega & \ldots & x_{2p_i}^{(i)} \\ \ldots & \ldots & \ldots & \ldots \\ x_{p_i1}^{(i)} & x_{p_i2}^{(i)} & \ldots & x_{p_ip_i}^{(i)} - \omega \end{pmatrix} \quad (90),$$

where O is a p_i-rowed square matrix all of whose elements are zero. The schematic determinant is p_j-rowed, and hence equals $| M_i |^{p_j}$. Thus, since there is a single unit E_i of character (i, i),

$$\delta = \prod_{i=1}^{H} | M_i |^{p_i + N_{i1}p_1 + N_{i2}p_2 + \ldots + N_{iH}p_H} \quad (91).$$

In particular, the characteristic determinant of the matric sub-algebra of p_i^2 units satisfying (87) equals $| M_i |^{p_i}$. Its rank equation is $| M_i | = 0$, as follows from Cayley's theorem on matrices (§ 15).

A determinant like $| M_i |$ whose elements are independent variables is an irreducible function of its elements.

The characteristic determinant for an algebra A_2 of the second category is a product of powers of H irreducible functions P_i of degrees p_i expressible as p_i-rowed determinants; and the Σp_i^2 elements of these H determinants are linearly independent linear functions of the coordinates. Here H is the number of partial moduli of the sub-algebra A_1.

If we equate to zero each element of these determinants, we obtain a system of linear equations which is independent of the choice of the units (§ 16). Hence the set of nilpotent numbers whose coordinates satisfy these equations is independent of the choice of units. When the normalized units of § 47 are used, these nilpotent numbers are the linear combinations of the η's.

49. Invariant sub-algebra; simple and semi-simple algebras.

An *invariant* sub-algebra I of any algebra A is one such that the product of any number of I by any number of A is in I and the product of any number of A by any number of I is in I.

For example, the algebra (e_1, e_2, e_3) in (46) has the invariant sub-algebras (e_1, e_2), (e_2) and (e_3). Algebra (45) has the invariant sub-algebras (e_i) and (e_i, e_j), $i, j = 1, 2, 3$.

A *simple* algebra is one having no invariant sub-algebra.

A *semi-simple** algebra is one having no nilpotent invariant sub-algebra. Algebra (45) is semi-simple since it has no nilpotent number, as shown by its characteristic equation, given after (45).

50. Main theorem. By § 37, the nilpotent numbers of an algebra A_1 of the first category form an invariant nilpotent sub-algebra. Hence A_1 is semi-simple if and only if its units are the partial moduli $\epsilon_1, \ldots, \epsilon_h$, and is then the direct sum of the unary algebras $(\epsilon_1), \ldots, (\epsilon_h)$. In particular, the only simple A_1 is that with a single unit ϵ_1.

In an algebra A_2 of the second category, the η's are, by (89), the units of an algebra which is nilpotent (end of § 48), and an invariant sub-algebra of A_2, by (88). Hence in a semi-simple A_2 the units η are absent, so that its units are those contained in the matric sub-algebras of p_i^2 units. As each of the latter is an invariant sub-algebra, a simple A_2 must be matric.

Conversely, any matric algebra M of p^2 units e_{ij} is simple. For, if $x = \Sigma x_{ij} e_{ij} \neq 0$ occurs in an invariant sub-algebra I, then

$$e_{\alpha i} x e_{j\beta} = x_{ij} e_{\alpha\beta}$$

occurs in I. Hence every $e_{\alpha\beta}$ is in I and $M \equiv I$.

Similarly, the direct sum $m + M$ of two matric algebras has no invariant sub-algebra I other than m and M. For, if $x + X$ be one of its numbers and $x \neq 0$, and if the units of m are e_{ij}, then

$$e_{\alpha i}(x + X) e_{j\beta} = x_{ij} e_{\alpha\beta}$$

is in I, so that m is a sub-algebra of I. Thus $I = m$ or $m + M$. If there be three or more summands, the argument is similar. Since a matric algebra or a direct sum of matric algebras is not nilpotent, a direct sum of matric algebras is semi-simple.

An algebra $(a_1, \ldots, a_r, b_1, \ldots, b_s)$ for which $A = (a_1, \ldots, a_r)$ and $B = (b_1, \ldots, b_s)$ are sub-algebras is called the *sum* of the algebras A and B. It need not be their direct sum (§ 21).

THEOREM†. *Any linear associative algebra with a modulus and*

* In place of Cartan's definition (*l.c.*, p. 57) of a semi-simple algebra as a reducible algebra each of whose irreducible components is simple, I have employed that by J. H. M. Wedderburn, *Proc. London Math. Soc.*, ser. 2, vol. 6 (1907), p. 94, and deduced the former property.

† Cartan, *l.c.*, p. 58. For simple algebras the theorem was first proved (but not altogether satisfactorily) by Th. Molien, *Math. Ann.*, vol. 41 (1893), p. 125; then by Frobenius, *Berlin Sitzungsber.*, 1903, p. 527; cf. J. B. Shaw, *Trans. Amer.*

with complex coordinates is the sum of a semi-simple algebra and a nilpotent invariant sub-algebra. A semi-simple algebra is either simple or a direct sum of simple algebras, and conversely. A simple algebra is of matric type and conversely.

51. Commutative algebras. A commutative complex algebra with a modulus must be of the first category (§ 46, Corollary). Employing the normalized units of § 39, we see that there cannot occur a unit η of character (α, β), $\beta \neq \alpha$. For, if so, $\epsilon_\alpha \eta = \eta$, $\eta \epsilon_\alpha = 0$, and multiplication would not be commutative. Then the algebra is the direct sum of $\Sigma_1, \ldots, \Sigma_h$, where Σ_j is formed of the numbers of character (j, j), and has r units $e = \epsilon_1, \eta_1, \eta_2, \ldots, \eta_{r-1}$ such that

$$e^2 = e, \quad e\eta_i = \eta_i e = \eta_i, \quad \eta_i \eta_j = \Sigma \gamma_{ijk} \eta_k \qquad (92),$$

where $k > i$, $k > j$, and $\gamma_{ijk} = \gamma_{jik}$. The γ's are of course subject to the further conditions implied by the associative law of multiplication.

In particular, if the algebra has no nilpotent number, no units η occur (for, $\eta^2_{r-1} = 0$), and the algebra is $(\epsilon_1, \ldots, \epsilon_h)$:

$$\epsilon_i^2 = \epsilon_i, \quad \epsilon_i \epsilon_j = 0 \quad (i, j = 1, \ldots, h\,;\ i \neq j) \qquad (93).$$

Math. Soc., vol. 4 (1903), pp. 275—283. It is implied in Frobenius's theory (see § 54) of group characters and group determinants, *Berlin Sitz.*, 1896, pp. 985, 1343; 1897, p. 994; 1898, p. 501; 1899, pp. 330, 482; 1904, p. 558, an elementary exposition of which was given by Dickson, *Annals of Math.*, ser. 2, vol. 4 (1902), pp. 25—49. For an attractive method of treating the latter theory, see I. Schur, *Berlin Sitz.*, 1905, p. 406 [cf. Dickson, *Trans. Amer. Math. Soc.*, vol. 8 (1907), p. 389]. Cf. W. Burnside, *Proc. London Math. Soc.*, vol. 35 (1903), p. 206. For the work of Wedderburn, see § 56. Another proof of the main theorem and the theorems in §§ 47, 48 has been given by Frobenius, *Berlin Sitz.*, 1903, p. 641.

PART III

RELATIONS OF LINEAR ALGEBRAS TO OTHER SUBJECTS

52. Correspondence between linear associative algebras and linear groups.

Consider a linear associative algebra with the multiplication table

$$e_i e_j = \sum_{s=1}^{n} \gamma_{ijs} e_s \qquad (i, j = 1, \ldots, n) \tag{94}$$

For y a fixed number and x a variable number of the algebra, the equation $x' = xy$ is equivalent to the n equations

$$(y) \qquad x_s' = \sum_{i,j=1}^{n} \gamma_{ijs} x_i y_j \qquad (s = 1, \ldots, n),$$

which define a linear homogeneous transformation of the variables $x_1, \ldots, x_n$ into the variables $x_1', \ldots, x_n'$, which corresponds to the given number y.

Similarly, to y' corresponds the transformation $x'' = x'y'$:

$$(y') \qquad x_t'' = \sum_{s,k=1}^{n} \gamma_{skt} x_s' y_k' \qquad (t = 1, \ldots, n)$$

of the variables $x_1', \ldots, x_n'$ into $x_1'', \ldots, x_n''$. But

$$x'' = (xy)\, y' = xy'', \quad y'' \equiv yy'.$$

To y'' corresponds the transformation

$$(y'') \qquad x_t'' = \sum_{i,s=1}^{n} \gamma_{ist} x_i y_s'', \; y_s'' \equiv \sum_{j,k=1}^{n} \gamma_{jks} y_j y_k' \qquad (t, s = 1, \ldots, n),$$

the values of the y_s'' being found from $y'' = yy'$ by use of (94). The transformation (y'') is therefore identical with the so-called *product*

$$(y)\,(y') \qquad x_t'' = \sum_{s,k,i,j=1}^{n} \gamma_{ijs} \gamma_{skt} x_i y_j y_k' \qquad (t = 1, \ldots, n),$$

obtained by eliminating the variables x_s' between the sets of equations

(y), (y'). To give a formal verification, note that the coefficients of $x_i y_j y_k'$ in the last two expressions for x_t'' are equal if

$$\sum_{s=1}^{n} \gamma_{ist}\gamma_{jks} = \sum_{s=1}^{n} \gamma_{ijs}\gamma_{skt} \qquad (i, j, k, t = 1, \ldots, n).$$

These are satisfied since, by (94),

$$e_i(e_j e_k) = e_i \sum_{s=1}^{n} \gamma_{jks} e_s = \sum_{t, s=1}^{n} \gamma_{ist}\gamma_{jks} e_t,$$

$$(e_i e_j)\, e_k = \left(\sum_{s=1}^{n} \gamma_{ijs} e_s\right) e_k = \sum_{s, t=1}^{n} \gamma_{ijs}\gamma_{skt} e_t.$$

Hence the correspondence between numbers y and transformations (y) is such that to the product yy' of two numbers corresponds the product $(y)(y')$ of the corresponding transformations. The resulting set of transformations is closed under multiplication since the algebra is.

For example, if the algebra is that of four units in § 4, to the number μ in (16), with the coordinates α, β, γ, δ, corresponds the transformation $m' = m\mu$ in (17_2):

$$a' = a\alpha + b\gamma, \quad b' = a\beta + b\delta, \quad c' = c\alpha + d\gamma, \quad d' = c\beta + d\delta \quad (95).$$

The product* of this transformation by that with the coefficients $\alpha', \ldots, \delta'$ is the transformation with the coefficients

$$\alpha'' = \alpha'\alpha + \gamma'\beta, \quad \gamma'' = \alpha'\gamma + \gamma'\delta, \quad \beta'' = \beta'\alpha + \delta'\beta, \quad \delta'' = \beta'\gamma + \delta'\delta \quad (96).$$

Let the algebra have a modulus ϵ. The corresponding transformation $x' = x\epsilon = x$ is the identity transformation $x_i' = x_i$ $(i = 1, \ldots, n)$. Then $\Delta(y)$ is not identically zero. For a number y such that $\Delta(y) \neq 0$, there exists a unique number y^{-1} such that $yy^{-1} = y^{-1}y = \epsilon$ (end of § 7). Then $x' = xy$ implies $x = x'y^{-1}$, so that the transformation (y^{-1}) is the inverse to (y) and conversely. If $\Delta(y) \neq 0$ and $\Delta(y') \neq 0$, then $\Delta(yy') \neq 0$, since yy' has the inverse $y'^{-1}y^{-1}$.

A set of linear transformations forms a *group* if the product of

* Since c, d are transformed cogrediently with a, b, the work need not be duplicated. If the matrix of the coefficients of a and b in (95) be called μ, we have $\mu'' = \mu'\mu$. The matrix of the coefficients of the product TT' of two linear transformations in n variables equals the product of the matrices of T' and T (i.e., taken in reverse order). But if we denote by T_1 the transformation

$$a = a'\alpha + b'\gamma, \quad b = a'\beta + b'\delta$$

of matrix μ, and by T_1' the transformation of matrix μ' expressing a', b' in terms of a'', b'', then by eliminating a', b', we obtain as the transformation expressing a, b in terms of a'', b'' one with the matrix $\mu\mu'$. Thus transformations, expressing the old variables in terms of the new, compound as their matrices taken in the same order. See also the second foot-note in § 53.

any two of the set is a transformation of the set, if the set contains the identity transformation, and if each transformation has an inverse in the set. Thus we have

THEOREM 1*. *The set of all numbers not nilfactors of a linear associative algebra with n units and having a modulus corresponds to a group G of linear transformations on n variables whose coefficients are linear functions of n arbitrary parameters.*

The determinant $\Delta'(y)$ of each transformation (y) of G is not zero. The group is called *simply transitive* since it contains a unique transformation (y) which replaces a given general set of values $x_1, \ldots, x_n$, viz., one for which $\Delta(x) \neq 0$, by any given set $x_1', \ldots, x_n'$ for which $\Delta(x') \neq 0$. Indeed, $xy = x'$ then has a unique solution y for which $\Delta(y) \neq 0$ and hence $\Delta'(y) \neq 0$.

To any number y corresponds also a transformation $x' = yx$:

$$[y] \qquad x_s' = \sum_{i,j=1}^{n} \gamma_{jis} x_i y_j \qquad (s = 1, \ldots, n).$$

Similarly, to y' corresponds $[y']$: $x'' = y'x'$. Then

$$x'' = y'(yx) = y''x, \quad y'' = y'y, \quad [y''] = [y'][y].$$

Hence we obtain a second simply transitive group G'.

If we apply a transformation $(y): x' = xy$ and then a transformation $[z]: x'' = zx'$, we obtain $(y)[z]: x'' = zxy$. The same result is obtained by applying first $[z]: x' = zx$ and afterwards $(y): x'' = x'y$. Hence *every transformation of G is commutative with every transformation of G'*. Two such simply transitive mutually commutative groups are called *reciprocal.*

The coefficients of the transformation, written at the end of formula (y''), replacing the variables $y_1, \ldots, y_n$ by the variables $y_1'', \ldots, y_n''$, are the same functions of the parameters $y_1', \ldots, y_n'$ as the coefficients of the x_i in (y) are of the parameters $y_1, \ldots, y_n$, as is also clear from the formulas $y'' = yy'$, $x' = xy$. For this reason, the group G is said to be its own *parameter group.* Similarly, $y'' = y'y$, $x' = yx$ show that G' is its own parameter group.

THEOREM 2†. *Every linear associative algebra with a modulus*

* Stated by Poincaré, *Paris Comptes Rendus*, vol. 99 (1884), p. 740.

† E. Study, *Leipzig Berichte*, vol. 41 (1889), math., p. 202; reprinted in *Monatshefte Math. u. Phys.*, vol. 1 (1890), pp. 283—355. While Study used the theory of bilinear forms, Scheffers gave a purely group-theoretic proof, using infinitesimal transformations (Lie-Scheffers, *Continuierliche Gruppen*, 1893, p. 627). Cf. Scheffers, *Leip. Ber.*, vol. 41 (1889), math., pp. 290—307.

defines a definite pair of reciprocal groups, each its own parameter group. Conversely, given any pair of simply transitive groups, reciprocal to each other, of linear homogeneous transformations on n variables, we can find new variables $x_1, \ldots, x_n$ *such that the groups are reduced simultaneously to groups of transformations* (y) *and* $[y]$ *in which the n parameters* $y_1, \ldots, y_n$ *enter linearly and homogeneously and such that each new group is its own parameter group; hence* (y) *defines a rule of multiplication* $x' = xy$ *of two numbers of a linear associative algebra with the units* $e_1, \ldots, e_n$ *and multiplication table* (94).

Consider the linear transformation $Q = q_1 q q_2$, where

$$q = x + yi + zj + wk, \quad Q = X + Yi + Zj + Wk$$

are quaternions with variable coordinates and q_1, q_2 are given quaternions. Then, by § 12,

$$N(Q) = X^2 + Y^2 + Z^2 + W^2 = c\,(x^2 + y^2 + z^2 + w^2), \quad c = N(q_1 q_2).$$

For $c \neq 0$, we have a transformation, involving seven arbitrary constants, expressing X, Y, Z, W as linear functions of determinant $\neq 0$ of x, y, z, w, which multiplies $x^2 + y^2 + z^2 + w^2$ by c. If we start with the general quaternary linear transformation having 16 coefficients and impose the last condition, viz., that the coefficients of the six terms XY, ... shall vanish, and those of X^2, ... shall be equal, we have nine relations and hence seven free constants. Supplementing this "counting of constants" by the definitive formula* for the resulting transformations, we see that all are given by the quaternion transformation $Q = q_1 q q_2$. We therefore have a very convenient expression for the group† generated by rotations around the origin and stretchings from it. To obtain the corresponding group‡ in three dimensions, we have only to take $x = X = 0$, $q_2 = q_1'$, the quaternion conjugate to q_1.

For further relations between linear algebras and groups of transformations not necessarily linear, and for various geometric aspects of our subject, the reader is referred to the *Encyc. Sc. Math.*, vol. I, 1,

* Cayley, 1854, *Coll. Math. Papers*, II, pp. 133, 214; Klein, *Math. Ann.*, vol. 37 (1890), p. 544, and *Elementarmath. vom höheren Standpunkte aus*, I, p. 161.

† Since if also $Q_1 = q_3 Q q_4$, then $Q_1 = (q_3 q_1)\, q\, (q_2 q_4)$.

‡ By retaining only transformations of positive determinants, we exclude

$$y' = -y, \quad z' = -z, \quad w' = -w$$

and its products by the former.

pp. 448—465. Also to p. 468 for the connection between the continuous groups of Sophus Lie and non-associative linear algebras in which $ab = -ba$, $(ab)c + (bc)a + (ca)b = 0$, for any three numbers a, b, c of the algebra.

53. Correspondence between linear associative algebras and systems of bilinear forms.

If x_i, u_i $(i = 1, \ldots, n)$ are $2n$ independent variables,

$$A = \sum_{i,j=1}^{n} a_{ij} x_i u_j$$

is called a *bilinear form*. Let B denote a similar form with the coefficients b_{ij}. Then $A + B$ denotes the form with the coefficients $a_{ij} + b_{ij}$. The product* AB is the bilinear form C with the coefficients

$$c_{ij} = \sum_{k=1}^{n} a_{ik} b_{kj} \qquad (i, j = 1, \ldots, n).$$

Thus the matrix of the coefficients of the sum or product of two bilinear forms is the sum or product of their matrices, taken in the same order. The subject of bilinear forms is therefore essentially identical with that of matrices and, if we agree to use also singular transformations (of determinant zero), with that of linear homogeneous transformations.

If at the beginning of § 52 we take $y = e_j$, the equation $x' = xe_j$ gives

$$x_s' = \sum_{i=1}^{n} \gamma_{ijs} x_i \qquad (s = 1, \ldots, n),$$

the matrix of which has the element γ_{ijs} in the sth row and ith column. Taking† it as the a_{is} in A, we obtain the bilinear form

$$A_j = \sum_{i,s=1}^{n} \gamma_{ijs} x_i u_s.$$

Then the number $\Sigma c_j e_j$ corresponds to the bilinear form $\Sigma c_j A_j$, while

* Hence $AB = \sum_k \frac{\partial A}{\partial u_k} \frac{\partial B}{\partial x_k}$. This is the definition by Frobenius, *Jour. der Math.*, vol. 84 (1877), pp. 1—63, the most important paper on the subject.

† If $TT' = T''$ for linear transformations then $\bar{\mu}\bar{\mu}' = \bar{\mu}''$, where $\bar{\mu}$ denotes the matrix μ of T with the rows and columns interchanged. Thus in (95), (96),

$$\begin{pmatrix} \alpha & \beta \\ \gamma & \delta \end{pmatrix} \begin{pmatrix} \alpha' & \beta' \\ \gamma' & \delta' \end{pmatrix} = \begin{pmatrix} \alpha'' & \beta'' \\ \gamma'' & \delta'' \end{pmatrix}.$$

the sum or product of two numbers corresponds to the sum or product of the corresponding bilinear forms*.

For example, the units 1, i of the binary algebra of complex numbers correspond to $x_1u_1 + x_2u_2$ and $x_1u_2 - x_2u_1$.

An excellent example is furnished by the matric algebra of four units. Since the corresponding group (95) affects the two pairs of variables cogrediently, the bilinear forms are each the sum of two similar parts on separate variables, x_iu_j, $i, j=1, 2$ in one part and $i, j=3, 4$ in the other part. Retaining only the first parts, we obtain

$$\alpha x_1u_1+\beta x_1u_2+\gamma x_2u_1+\delta x_2u_2$$

as the part corresponding to the transposed matrix of coefficients of a and b in (95). Thus each unit e_{ij} of the algebra [see (16_2)] corresponds to the bilinear form $B_{ij} \equiv x_iu_j$ with the same subscripts. The† B_{ij} satisfy relations (14).

Conversely, given n bilinear forms $A_1, \ldots, A_n$ in x_i, u_i $(i = 1, \ldots, n)$ such that each A_rA_s is a linear function of $A_1, \ldots, A_n$ with constant coefficients, we may take them as the units of a linear associative algebra.

The subject of bilinear forms‡ is therefore at bottom identical with that of linear associative algebras.

54. Relations of linear algebras to finite groups. Consider for example the rotations I, A, B in a plane about a point through angles 0°, 120°, 240°, respectively. They form a group with the multiplication table

	I	A	B
I	I	A	B
B	B	I	A
A	A	B	I

This is read $AB=I$, etc. Taking I, A, B as units whose products are given by the table, we have a linear associative algebra§. Similarly, if we start with any finite group, which is associative by definition.

* C. S. Peirce proved this theorem expressed in a different notation; see list of his papers in *Encyc. Sc. Math.*, vol. I, 1, p. 416. Cf. Study, *Leip. Ber.*, vol. 41 (1889), p. 192.

† The easiest verification is by differentiation (see first foot-note of this section).

‡ References are given in *Encyc. Sc. Math.*, vol. I, 1, pp. 415—421, 444—446. For an account of the theory of bilinear forms and their applications, see Muth, *Elementartheiler*, Leipzig, 1899.

§ Cayley, *Phil. Mag.*, vol. 7 (1854), p. 46 (=*Coll. Math. Papers*, vol. II, 1889, p. 129).

If we take I, A, B as ordinary variables, and consider the three-rowed matrix forming the body of the multiplication table of the above group (the multipliers having been arranged so that every element of the main diagonal is I), we have an example of what Frobenius* calls the *group matrix* of a given finite group. The product of two such group matrices is another of the same kind:

$$\begin{pmatrix} I & A & B \\ B & I & A \\ A & B & I \end{pmatrix}\begin{pmatrix} i & a & b \\ b & i & a \\ a & b & i \end{pmatrix} = \begin{pmatrix} \iota & \alpha & \beta \\ \beta & \iota & \alpha \\ \alpha & \beta & \iota \end{pmatrix}, \quad \begin{aligned} \iota &= Ii + Ab + Ba, \\ \alpha &= Ia + Ai + Bb, \\ \beta &= Ib + Aa + Bi. \end{aligned}$$

Call e_1 the matrix obtained from the first by setting $I = 1$, $A = B = 0$; e_2 that by setting $A = 1$, $I = B = 0$; e_3 by $B = 1$, $I = A = 0$. Then

$$(Ie_1 + Ae_2 + Be_3)(ie_1 + ae_2 + be_3) = \iota e_1 + \alpha e_2 + \beta e_3$$

is the equivalent of the above relation between matrices, and is the multiplication formula for the algebra with modulus e_1 and

$$e_2^2 = e_3, \quad e_2 e_3 = e_3 e_2 = e_1, \quad e_3^2 = e_2.$$

We have in effect returned to our group of three operators. For an elegant exposition of this direct relation between the theory of group matrices and linear algebras, and applications to Abelian integrals, see a memoir by Poincaré †.

55. Dedekind's point of view for linear associative commutative algebras ‡. Consider the example of the commutative triple algebra:

$$e_1^2 = e_1 + e_2 + e_3, \quad e_2 e_3 = e_2, \quad e_2^2 = e_3^2 = e_3, \quad e_1 e_2 = e_1 e_3 = e_2 + e_3.$$

Regarding these as ordinary algebraic equations for the unknowns e_i, we find at once that they have only four sets of solutions:

$$(e_1, e_2, e_3) = (0, 0, 0), \quad (1, 0, 0), \quad (2, 1, 1), \quad (0, -1, 1).$$

Excluding of course the first set, Dedekind regarded the e's as multi-valued (but correlated) numbers, so that any linear function of them is three-valued. The correlation is here such that, when $e_2 = r$, then $e_1 = r + 1$, $e_3 = r^2$, where r is a three-value function for which $r^3 = r$. In general, the multiplication equations

$$e_i e_j - \sum_{k=1}^{n} \gamma_{ijk} e_k = 0 \qquad (i, j = 1, \ldots, n) \tag{94}$$

* References at the end of § 50.

† *Journ. de Math.*, ser. 5, vol. 9 (1903), p. 180.

‡ *Göttingen Nachr.*, 1885, p. 141; 1887, p. 1; reproduced by B. Berloty, *Thèse*, Paris, 1886.

of a commutative associative algebra have n sets of real or complex solutions of determinant* not zero if and only if

$$\left| \sum_{i,j=1}^{n} \gamma_{rsi}\gamma_{ijj} \right| \neq 0 \qquad (r, s=1, \ldots, n).$$

Not far removed from this view is that of Kronecker†, who obtained all commutative linear associative algebras by use of modular systems, the moduli being the left members of (94).

* Cf. J. Petersen, *Gött. Nachr.*, 1887, p. 489; G. Frobenius, *Berlin Sitzungsber.*, 1896, p. 601; D. Hilbert, *Gött. Nachr.*, 1896, p. 179 (by the theory of invariants).

† *Berlin Sitzungsber.*, 1888, pp. 429, 447, 557, 595, 983. See the clear abstract in *Encycl. Sc. Math.*, vol. I, 1, pp. 409—411.

PART IV

LINEAR ALGEBRAS OVER A FIELD F

56. By a *division algebra* shall be meant one in which both right-hand and left-hand division, except by zero, are possible and unique. The fundamental theorem of § 50 on complex algebras has been generalized by Wedderburn* as follows :

THEOREM 1. *Any linear associative algebra over a field F is the sum of a semi-simple algebra and a nilpotent invariant sub-algebra, each over F. A semi-simple algebra over F is either simple or the direct sum of simple algebras over F. Any simple algebra over F is the direct product of a division algebra and a simple matric algebra each over F, including the possibility that the modulus is the only unit of one factor.*

As the part relating to simple algebras is quite different from our former result for the case of the complex field, it will clarify the subject to have a proof of the converse :

THEOREM 2. *The direct product P of a division algebra D and a matric algebra M, each over F, is simple ; any number of P commutative with every number of P is in D.*

Denote the units of M by e_{pq} $(p, q = 1, \ldots, n)$, so that the e_{pp} are the partial moduli and $\epsilon = e_{11} + \ldots + e_{nn}$ is the modulus of M and hence of P. Let I be any invariant sub-algebra of P, and $i = \Sigma \delta_{rs} e_{rs}$ be any number of I, where δ_{rs} is in D. Since I is invariant, $e_{pp} i e_{qq}$ is in I. Allowing p and q to vary, we get a product $\neq 0$ if $i \neq 0$, since

$$\sum_{p,\,q=1}^{n} e_{pp} i e_{qq} = \sum_{p=1}^{n} e_{pp} i \epsilon = \epsilon i \epsilon = i.$$

Thus, for a certain pair of integers p and q,

$$\delta_{pq} e_{pq} = e_{pp} i e_{qq} \neq 0.$$

* *Proc. London Math. Soc.*, ser. 2, vol. 6 (1907), p. 109.

Thus I contains a number $\delta_1 e_{pq}$, where δ_1 is a number $\neq 0$ of D. If δ is any element of D, we can find a number δ_2 of D such that $\delta_2\delta_1 = \delta$. Hence I contains δe_{pq} and therefore also

$$\delta e_{pq} e_{qr} = \delta e_{pr}, \quad e_{sp}\delta e_{pr} = \delta e_{sp} e_{pr} = \delta e_{sr},$$

where r, s are any positive integers $\leqq n$. Hence $I \equiv P$.

Next, if x is any number of P commutative with every number of P, set $x = \Sigma \delta_{rs} e_{rs}$, where δ_{rs} is in D. Then

$$\sum_{r=1}^{n} \delta_{rp} e_{rq} = x e_{pq} = e_{pq} x = \sum_{s=1}^{n} \delta_{qs} e_{ps},$$

where p and q are arbitrary. Hence

$$\delta_{rp} = 0\,(r \neq p), \quad \delta_{pp} = \delta_{qq}, \quad x = \delta_{11}\Sigma e_{qq} = \delta_{11}\epsilon = \delta_{11},$$

so that x is in D. The case in which D has a single unit gives the

Corollary. In a matric algebra M with the modulus ϵ, the numbers $a\epsilon$ (a scalar) are the only ones commutative with every number of M. Hence (§ 52) the only linear transformations commutative with everyone are those multiplying each variable by the same constant.

When F is the field of all complex numbers, any number of a division algebra satisfies a linear equation (§ 11), so that the algebra is the field itself. Hence (in agreement with § 50), any simple algebra is matric.

When F is the field of all real numbers, the only division algebras are the three found in § 11. We have therefore

Theorem 3*. *Every real simple algebra is either a real matric algebra M with p^2 units e_{jk}, or the direct product of M and the algebra $(1, i)$ of ordinary complex numbers, or the direct product of M and the algebra of real quaternions.*

For the second type, the units are e_{jk}, $e'_{jk} = ie_{jk} = e_{jk}i$, where

$$e_{jk}e'_{lm} = e'_{jk}e_{lm} = \begin{matrix} 0 \\ e'_{jm} \end{matrix}, \quad e'_{jk}e'_{lm} = \begin{matrix} 0 & (k \neq l), \\ -e_{jm} & (k = l). \end{matrix}$$

By the above corollary, a commutative matric algebra has a single unit ϵ. Using the first parts of Theorem 1 and Theorem 3, we get

* Expressed and proved otherwise by Cartan (*l.c.*, pp. 61—72). He also proved the first part of Theorem 1 for real algebras. The method was to study the corresponding complex algebra with the same units. We conclude that a real division algebra is simple and hence by Theorem 3 of one of the three types in § 11.

THEOREM 4. *Any real commutative algebra without nilpotent numbers is the direct sum of certain algebras* (ϵ_i), *each equivalent to the algebra of real numbers, and certain algebras* (e_i, e_i'),

$$e_i^2 = e_i, \quad e_i'^2 = -e_i, \quad e_i e_i' = e_i' e_i = e_i',$$

each equivalent to the algebra of ordinary complex numbers.

We obtain a sub-algebra of the complex algebra (93) by setting

$$E_i, E_i' = \tfrac{1}{2}(e_i \pm \sqrt{-1}\, e_i'), \quad E^2 = E, \quad E'^2 = E', \quad EE' = E'E = 0.$$

57. The algebras of Weierstrass. If k is a nilfactor (i.e., a number for which $ky = 0$ has a solution $y \neq 0$), and σ is not, there is an infinitude of numbers x satisfying the equation $k\rho + k\sigma x = 0$. For, if y is one of the infinitude of y's satisfying $ky = 0$, it has the solution x where x is determined from $\rho + \sigma x = y$. Similarly, there are equations of any degree m, having each coefficient c_i a multiple of k, with an infinitude of roots, viz., $x = x_0 + yz$, where $\Sigma c_i x_0^i = 0$. Weierstrass* investigated real linear algebras in which multiplication is commutative and associative, with $\Delta(x)$ not identically zero (§ 6), and such that the only algebraic equations having an infinitude of roots are those in which all of the coefficients are multiples of one and the same nilfactor. By a rather long, but elegant analysis, he proved that such an algebra is equivalent under a real linear transformation of units to a direct sum of real unary and binary algebras (as given in the above Theorem 4), and conversely that the resulting algebra has the specified properties.

This result follows at once from Theorem 4; for, if the algebra had a nilpotent number η, so that $\eta^k = 0$, the equation $x^k = 0$ would have an infinity of roots $r\eta$ (r real) and yet its coefficients are not nilfactors. But Theorem 4 was based upon results not proved in this tract. We shall therefore give a direct proof (the second of Study's two proofs †) of the theorem of Weierstrass, in the equivalent form (known to the latter) that the corresponding complex algebra is of type (93). It suffices to prove that the only complex irreducible algebras of Weierstrass are those with a single unit ϵ. By the criterion of Scheffers (§ 21) an irreducible commutative algebra with a modulus ϵ has no further number e such that $e^2 = e$. Thus (§ 20) the rank equation is $(a - \lambda\epsilon)^r = 0$. If $r > 1$, $(a - \lambda\epsilon)^{r-1}$ is a root $\neq 0$ of $x^2 = 0$. Hence $r = 1$ and the only unit is ϵ.

* *Göttingen Nachr.*, 1884, pp. 395—419.

† *Göttingen Nachr.*, 1898, p. 1 [1889, pp. 262—5]. All except the last papers cited in § 55 deal with this subject.

58. Commutative non-associative division algebras. Let F be any field (except one with the modulus 2 in which therefore $2x$ counts as zero) for which there is an irreducible function

$$x^3 - Bx^2 - \beta x - b \qquad (B, \beta, b \text{ in } F).$$

Then division except by zero is always possible and unique in the algebra* with the units 1, i, j over F for which

$$i^2 = j, \quad ij = ji = b + \beta i + Bj, \quad j^2 = 4bB - \beta^2 - 8bi - 2\beta j \qquad (97).$$

If the final equation be replaced by

$$j^2 = i(ij) = bB + (b + \beta B) i + (\beta + B^2) j,$$

we obtain the field $F(i)$, the determinant $\Delta(x)$ of whose general number $x = r + si + tj$ vanishes if and only if r, s, t are all zero. Hence the latter property holds also for the function δ derived from $\Delta(x)$ by replacing r, s, t by $r + \beta t$, $s + 2Bt$, $-2t$ respectively. But δ is the determinant of x for algebra (97).

Of quite another character is the commutative division algebra† with $2n$ units J^r, IJ^r $(r = 0, 1, \ldots, n-1)$ over F, where J is a root of a uniserial Abelian equation

$$x^n - c_1 x^{n-1} + c_2 x^{n-2} - \ldots \pm c_n = 0,$$

i.e., an equation irreducible in F and having the roots

$$J, J' = \Theta(J), J'' = \Theta(J') = \Theta^2(J), \ldots, J^{(n-1)} = \Theta^{n-1}(J) \quad [\Theta^n(J) \equiv J],$$

where Θ is a polynomial with coefficients in F, with the further condition that c_n is not the square of a number of F. The general number of the algebra is $A + BI$, where A, B are polynomials in J with coefficients in F. Write $B' = B(J')$, $B'' = B(J'')$, etc. for the conjugates to $B(J)$. The multiplication table of the algebra is defined by

$$(A + BI)(X + YI) = R + SI, \quad R \equiv AX + B'Y'J, \quad S \equiv BX + AY \qquad (98).$$

The last two equations can be solved for X and Y in the field $F(J)$ provided A and B are not both zero. Eliminating X, we get

$$BB'Y'J - A^2 Y = C, \quad C \equiv BR - AS.$$

* Dickson, "On finite algebras," *Göttingen Nachrichten*, 1905, pp. 358—393; *Bull. Amer. Math. Soc.*, vol. 14 (1907—8), p. 169, where these algebras are proved to be the only ones with three units and division unique if F is a finite field, by use of a remarkable theorem on non-vanishing ternary cubic forms; *Trans. Amer. Math. Soc.*, vol. 7 (1906), pp. 370—390, where (97) is found by invariantive properties.

† Dickson, *Trans. Amer. Math. Soc.*, vol. 7 (1906), p. 514.

In C, C', ..., $C^{(n-1)}$ the determinant of the coefficients of Y, Y', ... equals

$$c_n B^2 B'^2 \ldots [B^{(n-1)}]^2 - A^2 A'^2 \ldots [A^{(n-1)}]^2 \neq 0,$$

and the resulting Y satisfies the preceding equation. Hence division except by zero is always possible and unique.

59. Linear associative division algebras. Let $\phi(x) = 0$ be any uniserial Abelian equation of degree r in a field F. Call its roots

$$i, \quad \Theta(i), \quad \Theta^2(i) = \Theta[\Theta(i)], \ldots, \quad \Theta^{r-1}(i) \qquad [\Theta^r(i) = i].$$

Then $i^s j^t$ $(s, t = 0, 1, \ldots, r-1)$ are the r^2 units of a linear associative algebra* over F having

$$\phi(i) = 0, \quad ji = \Theta(i) j, \quad j^r = g \qquad (99),$$

where g and the coefficients of the polynomials Θ and ϕ are in F. If $r = 2$, we may, without loss of generality, take $i^2 = c$, where c is in F but is not the square of a number of F. The linear functions of i form a field $F(i)$. The general number of the algebra is $N = \alpha + \beta j$, α and β in $F(i)$. If $\beta = 0$, N has the inverse α^{-1}, in $F(i)$. If $\beta \neq 0$, $N = \beta\nu$, where ν is of the form $\gamma + j$, and $\gamma = e + fi$. Write $\gamma' = e - fi$. Since $\Theta(i) = -i$, we have

$$ji = -ij, \quad j\gamma = \gamma' j, \quad (j - \gamma')(j + \gamma) = g - \gamma\gamma'.$$

Hence† every number $N \neq 0$ has an inverse if g is not the norm $\gamma\gamma' \equiv e^2 - cf^2$ of a number γ of $F(i)$. The conditions on c and g are satisfied when F is the field of reals by taking $c = g = -1$, and the algebra is then that of real quaternions.

A like result holds‡ for any r. By (99_2),

$$jf(i) = f(\Theta) j, \quad j^2 f(i) = f[\Theta^2(i)] j^2 \qquad (100),$$

where f is a polynomial with coefficients in F. We shall treat the typical case $r = 3$. We have at once

$$[j^2 + k(\Theta^2) j + k(\Theta) k(\Theta^2)][j - k(i)] = g - k(i) k(\Theta) k(\Theta^2),$$

where $k(\Theta^2)$ denotes $k[\Theta^2(i)]$. Assume that g is not the norm of a number of $F(i)$. Then every $j - k(i)$ has an inverse. It remains only to show that

$$z = j^2 + \alpha j + \beta$$

has an inverse, α and β being in $F(i)$. But

$$z[j - \alpha(\Theta)] = [\beta - \alpha(i)\alpha(\Theta^2)] j + g - \beta(i)\alpha(\Theta)$$

* Dickson, *Trans. Amer. Math. Soc.*, vol. 15 (1914), p. 31.

† Dickson, *l.c.*, and *Trans. Amer. Math. Soc.*, vol. 13 (1912), pp. 65, 66.

‡ Wedderburn, *Trans. Amer. Math. Soc.*, vol. 15 (1914), p. 162.

is not zero since g is not the norm of $\alpha(i)$; and each number of $F(i)$ and each linear function of j has an inverse.

The existence of these linear associative algebras in r^2 units in which both right- and left-hand division except by zero is always possible and unique is of decided interest and importance in view of the rôle of division algebras in the general theory of § 56.

On the contrary, if the field F has only a finite number of elements multiplication must be commutative in a linear associative division algebra, which is therefore a field*.

60. Finite associative division algebras†. Consider p^2 couples (x, y), where x, y are integers taken modulo p, p a prime > 2, such that

$$(a, c) + (x, y) = (a + x, c + y) \qquad (101),$$

$$(a, c) . (x, y) = (ax + \epsilon\nu cy, cx + \epsilon ay) \qquad (102).$$

Here ν is a fixed quadratic non-residue of p, while

$$\epsilon \equiv (a^2 - \nu c^2)^{\frac{p-1}{2}} \pmod{p}.$$

Right-hand and left-hand division except by $(0, 0)$ are each possible and unique. Multiplication is associative. The first distributive law (4_1) holds, but not (4_2).

For p a prime of the form $3l + 1$, define the sum of (a, b, c) and (x, y, z) as in (101), and their product to be

$$(ax + \epsilon\nu cy + \epsilon^2\nu bz,\ bx + \epsilon ay + \epsilon^2\nu cz,\ cx + \epsilon by + \epsilon^2 az),$$

where ν is a fixed cubic non-residue of p, and

$$\epsilon \equiv D^{\frac{p-1}{3}} \pmod{p}, \quad D = a^3 + \nu b^3 + \nu^2 c^3 - 3\nu abc.$$

Now D is divisible by p only when a, b, c are. Indeed, D is the determinant $\Delta(x)$ for the field algebra of the numbers

$$x = a + b\rho + c\rho^2, \quad \rho^3 \equiv \nu \pmod{p}.$$

Hence each kind of division except by $(0, 0, 0)$ is possible and unique. Multiplication is associative, and the first distributive law (4_1) holds. The generalization to n-tuples is immediate.

* Wedderburn, *Trans. Amer. Math. Soc.*, vol. 6 (1905), p. 349; Dickson, *Göttingen Nachr.*, 1905, p. 381.

† Dickson, *Göttingen Nachr.*, 1905, p. 358.

Each of the algebras (97) and (101)—(102) has been used to construct non-Desarguesian and non-Pascalian geometries*.

61. Statement of further results on general linear algebras.

Let A be a linear algebra, not necessarily associative, over a field F. Let I be any invariant sub-algebra of A. Let $i_1, \ldots, i_p$ be a complete set of linearly independent numbers of I with respect to the field F. Choose numbers $a_1, \ldots, a_q$ of A such that $i_1, \ldots, i_p, a_1, \ldots, a_q$ form a complete set of linearly independent numbers of A with respect to F. Let $i, i', \ldots$ denote linear functions of $i_1, \ldots, i_p$ with coefficients in F; and $a, a', \ldots$ linear functions of $a_1, \ldots, a_q$. Then ai' and ia' are numbers i'' and i''' of the invariant sub-algebra I. Hence

$$(a+i)(a'+i') = aa' + i'''', \quad i'''' = ii' + i'' + i'''.$$

Hence if we suppress the components $x_1 i_1, \ldots, x_p i_p$ in all numbers and products of numbers of A, we obtain a linear algebra which is said to *accompany*† A and to be *complementary* to I. It is conveniently denoted by $A - I$ and called a *difference algebra*. The number of its units is the excess of the number of units of A over the number of units of I. By regarding as identical those numbers of A which differ only by numbers of I, we obtain $A - I$. It is an associative algebra if A is.

If each A_r is a maximal invariant sub-algebra of A_{r-1}, then A_1, A_2, $A_3, \ldots$ is called a *composition series* of A_1. The difference algebras $A_1 - A_2$, $A_2 - A_3, \ldots$ are said to form a *difference series* of A_1. *Any two difference series of A_1 differ only in the arrangement of the terms of the series*‡. Each algebra of the difference series is simple. A like theorem holds when each A_r is the largest sub-algebra of A_{r-1} invariant in A_1, the resulting series being called a principal or chief difference series. These theorems are analogous to those relating to a composition series of a finite group.

Let A_1 be a reducible algebra and A_2 its largest sub-algebra such that A_1 is the direct sum (§ 21) of A_2 and another sub-algebra A_2'. If A_2 is reducible, let A_3 be its largest sub-algebra such that A_2 is the direct sum of A_3 and another sub-algebra of A_2, etc. Then the series

* Veblen and Wedderburn, *Trans. Amer. Math. Soc.*, vol. 8 (1907), p. 379.

† Th. Molien, *Math. Ann.*, vol. 41 (1893), p. 93, *begleitendes System*; G. Frobenius, *Berlin Sitzungsb.*, 1903, p. 523, *homomorphe Gruppe*.

‡ Wedderburn, *Proc. London Math. Soc.*, ser. 2, vol. 6 (1907), pp. 84, 110; Epsteen and Wedderburn, *Trans. Amer. Math. Soc.*, vol. 6 (1905), p. 176.

$A_1 - A_2$, $A_2 - A_3$, ... is called* a *reduction series* of A_1. *Any two reduction series of A_1 differ only in the arrangement of their terms.*

By use of the latter theorem, Wedderburn proved that *a linear associative algebra over a field can be expressed in one and but one way as the direct sum of irreducible algebras each having a modulus and an algebra without a modulus.* In particular†, an algebra with a modulus can be expressed as the direct sum of irreducible algebras each having a modulus. The rank equation of $s + S$ is the product‡ of the rank equations of s and S.

62. Analytic functions of hypercomplex numbers. B. Berloty in his thesis (Paris, 1886) extended the elements of the theory of functions of a complex variable to that of a number in an algebra of Weierstrass. The conditions for such an extension to a general algebra with a modulus have been treated by G. Scheffers§; let $f_1, \ldots, f_n$ be continuous functions of $x_1, \ldots, x_n$; in order that $f = \Sigma f_i e_i$ shall have a unique derivative independent of $dx_1, \ldots, dx_n$, multiplication must be commutative; in order that the derivatives and integrals of analytic functions as defined shall be analytic, multiplication must also be associative. A different extension was based by F. Hausdorff‖ upon the number of linearly independent expressions $\Sigma a_i x b_i$, where the a's and b's are fixed numbers of the algebra.

* Wedderburn, *l.c.*, pp. 86, 112; Epsteen, *Trans. Amer. Math. Soc.*, vol. 4 (1903), p. 444.

† G. Scheffers, *Math. Ann.*, vol. 41 (1893), p. 601, for the case of the field of complex numbers. His proof rests upon an incorrect inference (top of p. 603), mentioned in the second foot-note to § 39 in this tract.

‡ Study, *Monatshefte Math. u. Phys.*, vol. 2 (1891), p. 44; Scheffers, *Math. Ann.*, vol. 39 (1891), p. 319.

§ *Paris Comp. Rend.*, 116 (1893), pp. 1114, 1242; *Leipzig Berichte*, vol. 45 (1893), math., p. 828; vol. 46 (1894), p. 120.

‖ *Leipzig Ber.*, vol. 52 (1900), math., p. 45. Cf. L. Autonne, *Paris Comptes Rendus*, 142 (1906), p. 1183; *Journ. de Math.*, ser. 6, vol. 3 (1907), p. 53.

Printed in the United States
By Bookmasters